BIOCHEMICAL SOCIETY SYMPOSIA

No. 58

THE ARCHAEBACTERIA: *BIOCHEMISTRY AND BIOTECHNOLOGY*

BIOCHEMICAL SOCIETY SYMPOSIUM No. 58
held at Heriot-Watt University, Edinburgh, September 1991

The Archaebacteria: *Biochemistry and Biotechnology*

ORGANIZED AND EDITED BY
M. J. DANSON, D. W. HOUGH AND G. G. LUNT

PORTLAND PRESS
London and
Chapel Hill

Published by Portland Press, 59 Portland Place, London W1N 3AJ, U.K.
on behalf of the Biochemical Society
In North America orders should be sent to Portland Press Inc.,
P.O. Box 2191, Chapel Hill, NC 27515–2191, U.S.A.

ISBN 1 855 78 010 0   ISSN 0067–8694

British Library Cataloguing in Publication Data
A catalogue record for this book is available from the British Library

Printed in Great Britain by the University Press, Cambridge

Front cover: Archaebacterial citrate synthase crystals (M. J. Danson, D. W. Hough, R. Russell and R. Acharya, Bath).

Contents

Preface

The pioneering work of Carl Woese has broken down the confines of thinking that organisms are prokaryotes or eukaryotes. Rather, he suggested that there are three lines of evolutionary descent — eubacteria, eukaryotes and archaebacteria. Archaebacteria were so named because they all grow in extreme environments which resemble those thought to have existed during the early periods of life on earth.

During the 14 years since the original description of the Archaebacteria, their unique nature has been recognized and this is emphasized in the recent suggestion by Woese, Kandler and Wheelis that living organisms should be classified into three domains — Archaea, Bacteria and Eukarya. It is clear from Doolittle's opening contribution to the Symposium that this suggested reclassification remains controversial. However, the uniqueness of the Archaebacteria (Archaea) and their exploitation of extreme environments has stimulated worldwide scientific interest which forms the basis of this Symposium.

The first day of the Symposium concentrated on the fundamental scientific interest of these organisms, with contributions reflecting the unique features of their biochemistry and molecular biology. The second day was devoted to more applied aspects, with sessions on Protein Stability in Extreme Environments and Biotechnological Potential of the Archaebacteria. In the latter area, discussion ranged from methanogenesis in ruminants and the possible implications for global warming to sophisticated pattern recognition devices based on bacteriorhodopsins — clear evidence that the Archaebacteria may indeed have an impact on our environment that even yet is not fully appreciated.

There can be no doubt that Archaebacteria, with their uniquely stable cellular constituents, will play an increasingly important role in the biotechnological industry of the future. However, it was also clear from discussions at the Symposium that future commercial applications will depend on basic research, which is essential if we are to fully understand macromolecular stability in extreme environments.

MICHAEL J. DANSON
DAVID W. HOUGH
GEORGE G. LUNT

Bath

Abbreviations

APS	Sulphuric acid–phosphoric acid anhydride
BR	Bacteriorhodopsin
$C_{carb.}$	Carbonate carbon
CDR factor	Carbon dioxide reduction factor
CH_3-S-CoM	2-(methylthio)ethanesulphonic acid
CODH	Carbon monoxide dehydrogenase
$C_{org.}$	Organic carbon
DGA	Desulphated diglycosylarchaeol
DGC	Diglycosylcardarchaeol
DHA	Dihydroxyacetone
DPE	Distal promoter element
EF	Elongation factor
EOR	Enhanced (or tertiary) oil recovery
F_{420}	Factor 420
F_{430}	Factor 430
FAF	Formaldehyde activation factor
GAPDH	Glyceraldehyde-3-phosphate dehydrogenase
GNC	Glucosylnonitolcaldarchaeol
HB	3-Hydroxybutyrate
hMDH	Halophilic malate dehydrogenase
H_4MPT	Tetrahydromethanopterin
HS-CoM	2-Mercaptoethanesulphonic acid/Coenzyme M
HS-HTP	7-Mercaptoheptanoylthreonine phosphate
HV	3-Hydroxyvalerate
LDH	Lactate dehydrogenase
LS	Light scattering
M_2	Molar mass
MDH	Malate dehydrogenase
MPT	Methanopterin
MT	Methyltransferase
NS	Neutron scattering
ORF	Open reading frame
PA	Phosphatidic acid
PAPS	3′-Phosphoadenosine-5′-phosphosulphate
PCR	Polymerase chain reaction
PE	Phosphatidylethanolamine
PG	Phosphatidylglycerol
PGC	Phosphodiglycosylcaldarchaeol

PGNC	Glc-nonitolcaldarchaeol-*P*-inositol
PGP	Phosphatidylglycerol phosphate
PGS	Phosphatidylglycerol sulphate
PHA	Polyhydroxyalkanoate
PHB	Poly-3-hydroxybutyrate
PI	Phosphatidylinositol
PM	Purple membrane
PPE	Proximal promoter element
PS	Phosphatidylserine
RuBP	Ribulose bisphosphate carboxylase
S-DGA	Sulphated diglycosylarchaeol
S-TeGA	Sulphated tetraglycosylarchaeol
S-TGA	Sulphated triglycosylarchaeol
SD	Sedimentation velocity and diffusion
SE	Sedimentation equilibrium
TeGA	Desulphated tetraglycosylarchaeol
TGA	Desulphated triglycosylarchaeol
TMP	Trimethylpsoralen
$T_{opt.}$	Optimal growth temperature
TPP	Thiamine pyrophosphate
XS	X-ray scattering
YFC	Yellow fluorescent compound

Biochem. Soc. Symp. **58**, 1–6
Printed in Great Britain

What are the archaebacteria and why are they important?

W. Ford Doolittle

Department of Biochemistry, Dalhousie University, Halifax, Nova Scotia, Canada B3H 4H7

Purpose

My purpose here is to review changing concepts about the nature, number and naming of the major groups into which all living things can be divided, emphasizing the role played in that intellectual history by the archaebacteria, since their 'discovery' in 1977. I hope this will set the stage for subsequent chapters, in which the reasons for much of the current excitement in archaebacteriology are made apparent, and future directions mapped out. Otto Kandler has covered much of the same ground before, and takes a different direction in the concluding remarks published in this volume.

Experimental scientists often rewrite history quite shamelessly (see discussion by Sapp [1]). In our primary literature we present experiments in the order that makes our progress from preliminary observation through hypothesis to con firmation seem ineluctable and dramatic, often not in the haphazard way the work was actually done. More broadly, in presenting our results as part of the larger developing logic of our scientific disciplines, we chose accounts of that discipline which make our results seem especially interesting, sought-after or challenging. The literature of experiment and that of the history of our science are thus parallel and interactive but not always responsible to each other. I make no claim to have recreated the dialogue between them accurately in the narrative which follows.

Kingdoms of life

Those of us who received secondary or university educations in the first half of this century were taught that living things were either animals or plants, and that while fungi, protozoa and bacteria might be difficult to classify, they were best thought of as just very simple animals or plants. Life progressed throughout its history from simple to complex on the two parallel tracks, animal or plant; bacteria had gotten off the ladder early. Those educated more recently had to reconcile this view with the growing realization, then hardening into dogma, that "the line of

Table 1. Some features distinguishing prokaryotes from eukaryotes, as we knew them in 1977.

Prokaryotes	Eukaryotes
Circular chromosomes	Linear chromosomes
No histones	Histones
Single RNA polymerase	Three RNA polymerases
No caps or polyA on mRNA	Caps and polyA on mRNA
No introns	Introns[a]
Shine/Dalgarno sequence[a]	No Shine/Dalgarno sequence
Polycistronic mRNA	Monocistronic mRNA
70S ribosomes	80S ribosomes[a]
Murein walls[a]	No murein
Formyl methionyl initiator tRNA[a]	Unformylated initiatior tRNA

[a] Traits we could not include in such a table in 1991, because they are now known not to be universally (or nearly universally) present in one group and universally (or nearly universally) absent from the other. Most such losses in certainty come from investigations of the archaebacteria or protists diverging from below the trypanosomes.

demarcation between eukaryotic and prokaryotic cellular organisms is the largest and most profound single evolutionary discontinuity in the contemporary biological world" [2]. We saw tables, such as Table 1, listing the differences between eukaryotes and prokaryotes, and Stanier and van Niel could argue as soon as 1961 [3] that some of these traits, such as peptidoglycan or 70S ribosomes allowed us to define prokaryotes in a positive way, not simply as 'other than eukaryotes' because they lacked nuclei. (It is not clear, however, that an unambiguous definition of 'prokaryote' which includes eubacteria and archaebacteria, and is based on positive features, will be possible.)

Copeland and Whittaker provided grand views which attempted to honour both animal/plant and prokaryote/eukaryote dichotomies, while also addressing the difficulty of describing many eukaryotic unicells as unambiguously either animals or plants [4]. Copeland defined three eukaryotic kingdoms: Metaphyta, Metazoa and Protoctista, and one of the prokaryotes: Monera. Whittaker took the Fungi out of the Protoctists (renamed Protists) to grant them kingdom status too, and his five-kingdom system remains perhaps the standard view today. It suffers from still not doing enough justice to the eukaryote/prokaryote split: at the cellular and molecular level the changes wrought in the prokaryote/eukaryote (moneran/protist) transition were profoundly more radical (and difficult for us to rationalize) than any of the innovations or shifts in nutritional mode which form the basic distinctions between Whittaker's four kingdoms of eukaryotes.

The endosymbiont hypothesis

Margulis [5] helped much here, by arguing, in the late 1960s, that eukaryotic cells were in fact chimaerae—the results of the fusion of two, three or four prokaryotic lineages. The ultrastructural, physiological and genetic rearrangements required for that many genomes to function well as a team in a new unit of selection, the individual cell, must have been catastrophic and sweeping. Perhaps they were enough, in sum, to account for Table 1.

Molecular sequence data confirming at least parts of Margulis' version of the serial endosymbiosis theory began to appear in the 1960s and early 1970s. Sequencing of cytochromes and ferredoxins made the relationships between plastids and photosynthetic prokaryotes (specifically cyanobacteria) and between mitochondria and respiring bacteria like *Paracoccus* or *Rhodobacter* abundantly clear. The only extensive data tracking the nuclear genome of eukaryotic cells were 5S ribosomal RNA (rRNA) sequences.

Schwartz and Dayhoff, in 1978, put together a composite tree from all three data sets [6]. In this tree, the eukaryotic nuclear lineage appears to branch off from within the bacteria, between *Bacillus subtilis* and *Escherichia coli*. There were no other good reasons to support such a branching, and the 5S molecule (because of its small size and regions of strong conservation) began to seem a poor molecular chronometer, especially since the results of Woese's characterizations of the much more useful 16S rRNA had just begun to appear [7].

Comparisons of catalogues (lists of the sequences of T1 ribonuclease digestion products) of 16S rRNAs from bacteria, mitochondria and chloroplasts, and the homologous nucleus-encoded 18S rRNA, reconfirmed the purple bacterial and cyanobacterial origins of mitochondria and, more importantly, showed what cytochrome or ferredoxin sequences could not: that the nucleus was of some altogether different provenance. In fact, there was no bacterial lineage which could be shown to be closer to eukaryotic nuclei than any other. The two main branches of the tree of Life diverged at its root, Woese suggested, not two-thirds of the way up as common sense and the fossil record seemed to dictate.

If this were true, then the differences between eukaryotes and prokaryotes tabulated in Table 1 might not represent a radical reworking of cells and genomes at the time of some 'prokaryote/eukaryote transition'. They could instead indicate that the eukaryotic nuclear and prokaryotic lineages had separated at a time when cells and genomes were still very primitive, still 'in the throes of refining the genotype–phenotype coupling', and had since achieved different solutions to problems of accuracy and speed in information processing unsolved in the ancestor. (The fact that eukaryotic cells do not appear in the fossil record until relatively recently—1.5 billion years ago—need not be a problem, since extensive changes in cellular organization and size might have been caused by the acquisition of proto-organellar endosymbionts, as Margulis had said.)

Recognition of the archaebacteria

Woese's data provided a further surprise. There was a second group of prokaryotes as different from the eubacteria (from which almost all of our comparative biochemical and genetic information and all evolutionary models had been taken) as these were from eukaryotes. This group, the archaebacteria, commanded our attention because (i) studying their properties might help us deduce the properties of that primitive last common ancestor, which Woese called 'the progenote', and (ii) they were there, and likely to provide us with many new examples of how genes and genomes could be put together. Woese, Fox and collaborators [7] considered archaebacteria, eubacteria and eukaryotes (or the nuclear lineage thereof) as Life's three kingdoms, and the three-kingdom classification of these authors has replaced the five-kingdom scheme of Whittaker in some texts, particularly those written by molecular biologists and microbiologists.

Deduction of the properties of the progenote, determining which of those features which distinguish the three kingdoms today are primitive and which derived, requires sure knowledge of the branching order at the base of the tree of Life. For example, if archaebacteria and eukaryotes are sister groups, with eubacteria diverging earlier, then we can deduce whether the characteristic branched-chain ether lipids of archaebacteria or the straight-chain ester lipids of eukaryotes are the more recent invention by asking which kind of lipids eubacteria have. If they show straight-chain lipids (as they do) then it is most economical to assume this feature is primitive—we would otherwise have to assume extensive convergence in eukaryotic and eubacterial lineages. But if eubacteria and eukaryotes were sister groups, either lipid type might be primitive and parsimony could not tell us which.

Rooting the tree

Trees cannot be rooted without outgroups, and no tree relating all Life can have an outgroup. Schwartz and Dayhoff had used an internal duplication in the ferredoxin gene to get around this, in rooting their 1978 composite tree. Iwabe and coworkers [8] approached the rooting of the most recent universal tree similarly. All organisms contain pairs of genes for elongation factors EF-TU/1α and EF-G/2, so the duplication giving rise to this gene pair was already present in the last common ancestor of all Life. Thus, for a tree relating all EF-Tu sequences, any EF-G sequence is the functional equivalent of an outgroup: it diverges from below the last common ancestor. We can use EF-G trees to root organismal phylogenies based on EF-Tu and *vice versa*, and Iwabe *et al.* [8] identified another gene pair (α and β subunits of F1-ATPases) to which this logic could be applied. Both sets of analysis show archaebacteria and eukaryotes to be sister groups, with eubacteria most deeply diverged.

This conclusion seems to confirm the suspicions and hopes of many archaebacteriologists. The suspicions came from the recognition that a number of archaebacterial proteins, especially those involved in transcription and translation, resemble very strongly their eukaryotic homologues—in sequence, function and subunit interactions. The hopes have to do with the fact that archaebacteria become even more interesting if, in fact, they are the closest living relatives of the anucleate ancestor of nucleated cells, rather than just another, albeit very different, group of prokaryotes.

This conclusion then also brings with it a conundrum. What we know of gene organization in archaebacteria shows a very typically eubacterial pattern. Genes for the enzymes of tryptophan biosynthesis are organized into operons, although different in order from eubacterial tryptophan operons. Genes for RNA polymerase subunits are organized into operons showing eubacterial structure, however, and ribosomal protein gene operons show remarkable conservation in order, extending over dozens of genes, between *E. coli*, *B. subtilis*, *Methanococcus voltae* and *Halobacterium cutirubrum* [9]. A pattern of operon organization, and indeed certain specific operons themselves, must have already been established in the last common ancestor of archaebacteria and eubacteria, and thus in the last common ancestor of all Life. This means that eukaryotes must have lost operons in some radical genomic reorganization since their divergence from archaebacteria. Thus at least this characteristic difference between eukaryotes and 'prokaryotes' cannot be taken as reflecting different solutions to problems unsolved in their primitive common ancestor. Therefore we can no longer be so certain that the last common ancestor of all Life was really the primitive 'progenote'—the progenote could have become extinct aeons earlier.

Other solutions may be possible. Zillig *et al.*, in these proceedings (pp. 79–88), discuss the notion that the eukaryotic nucleus itself is completely chimaeric, with some genes (RNA polymerases II and III, for instance) of archaebacterial origin and others (RNA polymerase I) from eubacterial sources, and maybe none defining an ancient eukaryotic nuclear lineage. The idea is appealing, although one could by the same logic suggest that archaebacteria picked up many of their genes from the eukaryotic nuclear lineage, using them to replace, *in situ* and without disrupting gene organization, older, perhaps less 'highly evolved' and more eubacteria-like homologues.

Other cautions

The conclusions drawn by Iwabe *et al.* [8] are of immense importance, but based on limited data, and it may be too soon to take them so seriously. Forterre *et al.* argue elsewhere in this volume (pp. 99–112) that eubacteria and archaebacteria are sister groups, derived by streamlining an ancestral mesophilic eukaryote-like cell and genome to adapt to high temperature, and that the pattern of gene duplication assumed by Iwabe *et al.* is oversimple. There certainly are a number of protein gene primary sequences which belie the closer affinity between archaebacteria and eukaryotes, several presented in this symposium.

One further caution should be raised. Many of us have written about eubacterial, archaebacterial or eukaryotic features of gene and genome organization as if these were well defined and conserved within each group. The assumption ignores the fact that we have only a very limited range of comparative molecular biological data from within the eubacteria (we know nothing of *Thermotoga*, say), or from within the eukaryotes (the animals, plants and fungi from which almost all our knowledge derives are from but one small branch of the eukaryotic tree). Ironically, we have a broader comparative data base for archaebacteria. It may well be that full knowledge will reveal a complete spectrum of types of gene organization and mechanisms of expression, with no sharp boundaries in genomic 'style' between kingdoms. Many of us also assume a uniformity in gene structure and function within single genomes which the data do not fully support. There are within *E. coli*, for instance, two types of promoter structures and control systems. Collado-Vides *et al.* [10] have pointed out that the less-written about σ^{54}-dependent promoters behave more like eukaryotic RNA polymerase II promoters than like canonical σ^{70} 'Pribnow box' promoters.

Woese *et al.* [11] have argued from the data supporting the distinctness and monophyly of archaebacteria, eubacteria and eukaryotes that these are the major groupings of living things, and since eukaryotes have long been considered to comprise several 'kingdoms', the new larger assemblages should be called 'domains'. Support for three domains, not two (prokaryotes and eukaryotes) nevertheless comes in no small part from the Iwabe tree, which would not allow a monophyletic and holophyletic prokaryotic domain.

Margulis and Guerrero criticized the three domain view because it is based so largely on molecular data [12]. Their views point out the still profound disagreement among different kinds of biologists about what a phylogenetic taxonomy is. If birds and dinosaurs were shown by total genomic sequencing to differ by but a dozen base pairs, molecular biologists would claim (correctly, I believe) that the marvel thus revealed would be that so very few base changes can lead to such drastic pleiotropic alterations in developmental programmes. Biologists like Margulis would, however, claim (wrongly, I feel) that this showed once again that an entire genome sequence is just another character, no more important than feathers and scales or songs and mating behaviour in defining relationships. Mayr has made similar objections [13], holding that, after all, archaebacteria and eubacteria are really prokaryotes. He makes the mistake Woese and his collaborators have cautioned us against, and assumes what he claims to prove. The words 'prokaryote' and 'eukaryote' describe grades, not clades.

Woese *et al.* [11] name the three domains Archaea, Bacteria and Eukarya. I retain reservations about the adequacy of the data supporting the tree of Iwabe *et al.* [8], and fear that much of the biological community is still grappling with the basic issues the discovery of the archaebacteria first raised, and so have been personally slow to adopt these terms. As the papers in this collection show, however, the new language is spreading. There seems no reason to hold out much longer.

References

1. Sapp, J. (1990) Where the Truth Lies, Cambridge University Press, Cambridge
2. Stanier, R. Y., Adelberg, E. A. & Ingraham, J. (1976) The Microbial World, 4th edn, Prentice Hall, Engelwood Cliffs
3. Stanier, R. Y. & van Niel, C. B. (1962) Arch. Mikrobiologie **42**, 17–35
4. Whittaker, R. H. (1969) Science **163**, 150–162
5. Margulis, L. (1970) Origin of Eukaryotic Cells, Yale University Press, New Haven
6. Schwartz, R. M. & Dayhoff, M. O. (1978) Science **199**, 395–403
7. Fox, G. E., Stackebrandt, E., Hespell, R. B., Gibson, J., Maniloff, J., Dyer, T. A., Wolfe, R. S., Balche, W. E., Tanner, R. S., Magrum, L. J., Zablen, L. B., Blakemore, R., Gupta, R., Bonen, L., Lewis, B. J., Stahl, D. A., Luehrsen, K. R., Chen, K. N. & Woese, C. R. (1980) Science **209**, 457–463
8. Iwabe, N., Kuma, K. I., Hasegawa, M., Osawa, S. & Miyata, T. (1989) Proc. Natl. Acad. Sci. U.S.A. **86**, 9355–9359
9. Auer, J., Lechner, K. & Bock, A. (1989) Can. J. Microbiol. **35**, 200–204
10. Callado-Vides, J., Magasanik, B. & Gralla, J. D. (1991) Microbiol. Rev. **55**, 371–394
11. Woese, C. R., Kandler, O. & Wheelis, M. L. (1990) Proc. Natl. Acad. Sci. U.S.A. **87**, 4576–4579
12. Margulis, L. & Guerrero, R. (1991) New Sci. **129**, 456–50
13. Mayr, E. (1991) Nature (London) **348**, 491

Biochem. Soc. Symp. **58**, 7–21
Printed in Great Britain

The enzymology of archaebacterial pathways of central metabolism

Michael J. Danson and David W. Hough

Department of Biochemistry, University of Bath, Bath BA2 7AY, U.K.

Synopsis

From a comparison of the pathways of central metabolism in the archaebacteria, eubacteria and eukaryotes, it is clear that the basic pathways were established before the divergence of the three kingdoms, but that the notable differences may provide important clues to their evolution. From these comparisons, enzymes found in all evolutionary groups have been chosen for detailed structural studies; given the range of extreme phenotypes found within the archaebacteria, these studies will be crucial to our understanding of the structural basis for protein stability and how such features may be engineered into a protein of choice.

Introduction

The pathways of central metabolism provide the metabolic links between the catabolic (degradative) and anabolic (biosynthetic) routes in living organisms, and also serve as their major pathways of energy generation. Central metabolism can therefore be considered in three stages: the catabolism of hexose sugars to pyruvate, the conversion of pyruvate to acetyl-CoA, and the oxidation of acetyl-CoA via the citric acid cycle. The purpose of this review is to describe these pathways in the archaebacteria, and to compare them with those found in eubacterial and eukaryotic species. This description will be in the form of a brief outline, as we have recently reviewed archaebacterial metabolism in detail, together with the evidence for the proposed routes [1,2].

From these central pathways, several archaebacterial enzymes have been chosen for structure–function studies in our laboratory, and this work will also be reviewed, again in comparison with the eubacterial and eukaryotic counterparts. In the longer term, these enzymological studies will not only contribute valuable data to our understanding of the evolutionary relationships between the three kingdoms, but they will provide important clues to the structural basis of protein stability in extreme environments.

Throughout this paper, the archaebacteria are considered in terms of their phenotypes (halophiles, thermophiles, methanogens), although it is realized that these do not necessarily reflect genotypic relationships [3,4].

(I) Central Metabolism

Central metabolism in eubacteria and eukaryotes

The Embden–Meyerhof glycolytic pathway for the catabolism of glucose is characteristic of eukaryotes and many anaerobic eubacteria, and yields two ATP molecules per glucose metabolized. However, in many strictly aerobic eubacteria, the glycolytic enzyme, 6-phosphofructokinase, is absent and glucose is catabolized via the Entner–Doudoroff pathway with the yield of only one ATP per glucose. In both eubacteria and eukaryotes, the pentose phosphate cycle is also present for glucose metabolism; it serves mainly to generate NADPH and tetrose and pentose sugars, and therefore may be considered as a supplementary pathway to the two main catabolic routes. In all three pathways (Fig. 1), there is a common trunk sequence from triose phosphate to pyruvate and this may be considered as the true central pathway for sugar catabolism [5].

Under aerobic conditions, pyruvate can be oxidatively decarboxylated to acetyl-CoA via the pyruvate dehydrogenase complex [6], and then oxidized to CO_2 and H_2O via the citric acid cycle. In eubacteria growing anaerobically, pyruvate can be metabolized fermentatively [7], and thus may serve as an electron sink for reducing equivalents generated in its formation from glucose.

Hexose catabolism in the archaebacteria

ATP-phosphofructokinase has not been detected in any archaebacterial species, and therefore the glycolytic pathway would appear not to be used in its catabolic mode. Instead, variations on the Entner–Doudoroff pathway have been discovered.

In the extreme halophiles, glucose and galactose are catabolized via a semi-phosphorylated Entner–Doudoroff pathway (Fig. 2), where glucose is oxidized to gluconate and then dehydrated to 2-keto-3-deoxygluconate, before phosphorylation to give 2-keto-3-deoxy-6-phosphogluconate [8–11]. Metabolism then proceeds via the normal Entner–Doudoroff route, with the yield of one ATP per glucose catabolized.

The thermophilic archaebacteria, *Sulfolobus* and *Thermoplasma*, utilize a further modification of this pathway, in that the 2-keto-3-deoxygluconate undergoes direct aldol cleavage to pyruvate and glyceraldehyde without prior phosphorylation [2,12–14]. Glyceraldehyde is then oxidized to glycerate, which is converted into 2-phosphoglycerate by glycerate kinase, with enolase and pyruvate kinase completing the production of a second molecule of pyruvate (Fig. 2). No metabolic data are yet available for other thermophilic archaebacteria. However, the fermentative anaerobic archaebacterium, *Pyrococcus furiosus*, has been found to contain ferredoxin-linked glucose oxidoreductase (catalysing glucose to gluconate) and glyceraldehyde oxidoreductase (glyceraldehyde to glycerate), and a pathway similar to that in *Sulfolobus* and *Thermoplasma* has been suggested but one in which ferredoxin substitutes for $NAD(P)^+$ [15].

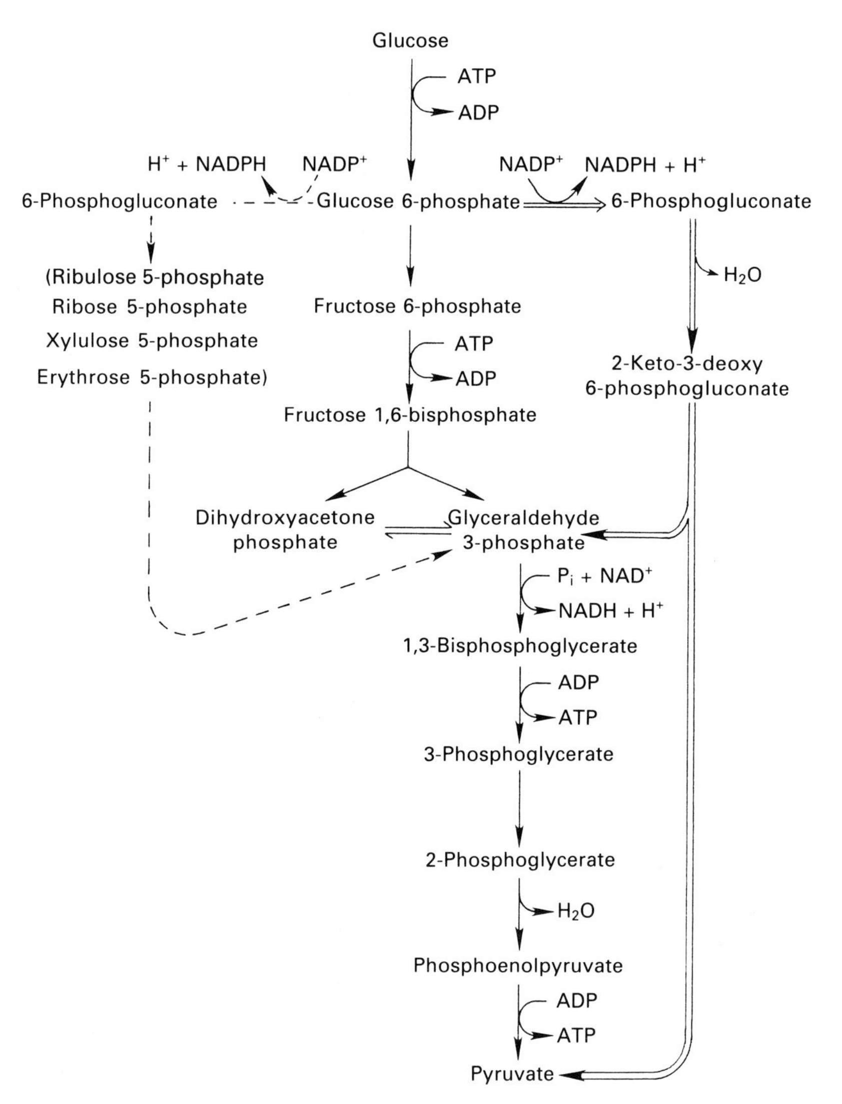

Fig. 1. Pathways of glucose catabolism in eubacteria and eukaryotes. The Embden–Meyerhof glycolytic pathway (→), the Entner–Doudoroff pathway (⇒) and the pentose phosphate pathway (- - - →).

There is no net yield of ATP in this non-phosphorylated Entner–Doudoroff pathway and, as described later, in the aerobic thermophiles *Thermoplasma* and *Sulfolobus*, energy is probably generated by an oxidative citric acid cycle. However, additionally in *Thermoplasma acidophilum* [2,16] and in *P. furiosus* [17], acetate can be generated from acetyl-CoA with the concomitant production of ATP via the enzyme acetyl-CoA synthase (ADP-utilizing):

$$\text{Acetyl-CoA} + \text{ADP} + \text{P}_i \rightarrow \text{Acetate} + \text{CoA} + \text{ATP}$$

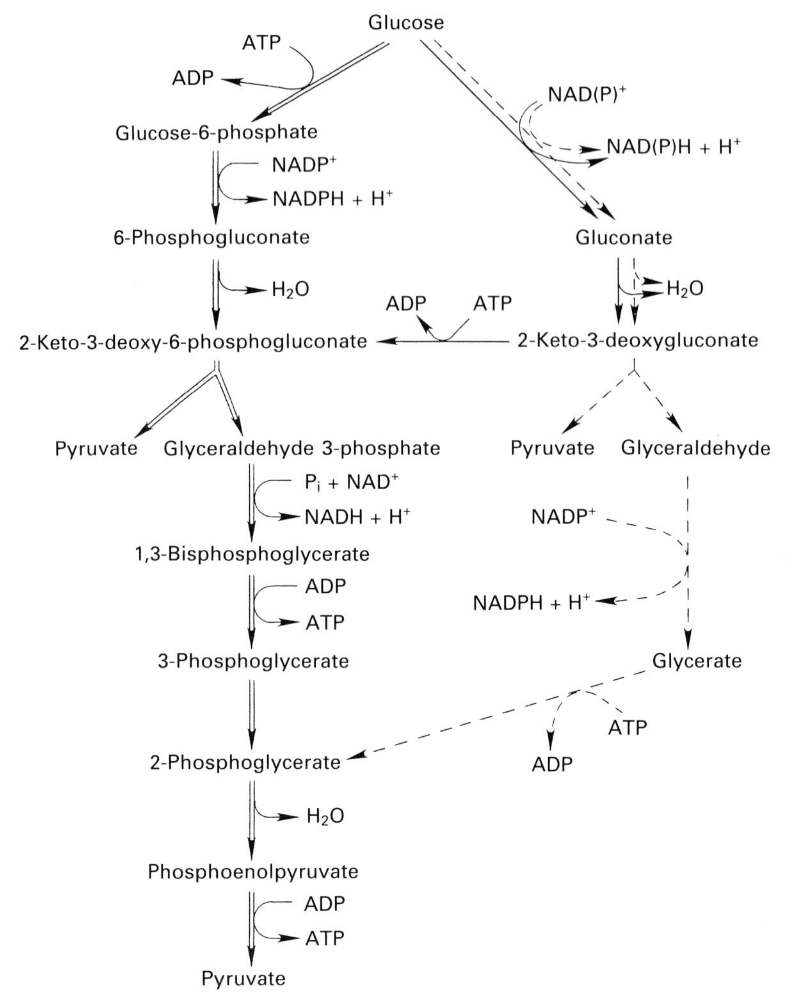

Fig. 2. Pathways of glucose catabolism in halophilic and thermophilic archaebacteria. The modified Entner–Doudoroff pathway in halophiles (→) and the non-phosphorylated Entner–Doudoroff pathway in *Sulfolobus solfataricus* and in *Thermoplasma acidophilum* (– – – →) are shown in comparison with the classical Entner–Doudoroff pathway (⇒) from Fig. 1.

Most methanogenic archaebacteria are autotrophs, generating energy by the formation of methane from CO_2 and H_2. Some are able to use C_1 compounds, but in all methanogens carbon is eventually fixed into acetyl-CoA. Therefore, carbohydrate synthesis, rather than catabolism, is the main direction of central metabolism, and there is extensive evidence that gluconeogenesis proceeds via a reversal of the Embden–Meyerhof pathway [18,19]. However, there must be some degree of carbohydrate turnover in these archaebacteria, and the evidence from ^{13}C n.m.r. spectroscopy suggests that *Methanobacterium thermoautotrophicum* can metabolize glucose to pyruvate via a normal glycolytic pathway [20,21]. At first sight, this

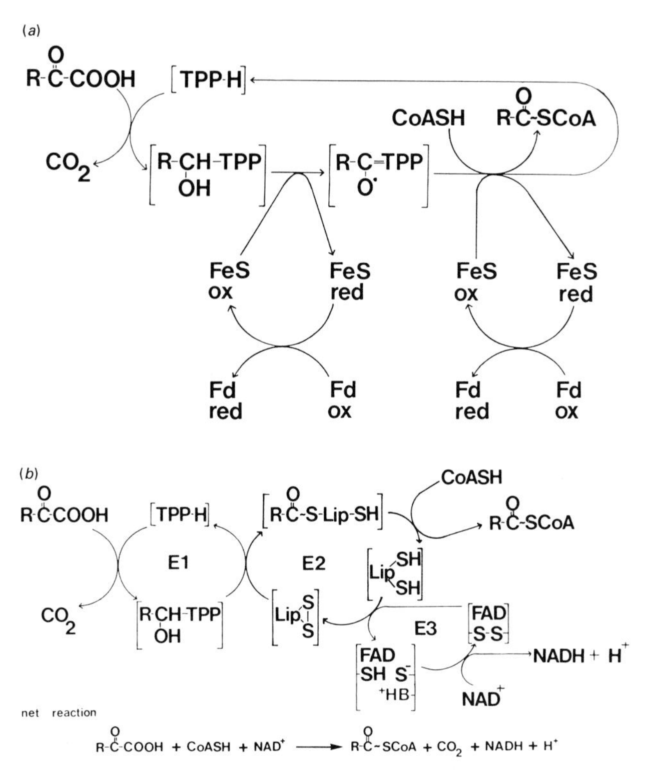

Fig. 3. Pyruvate oxidoreductase and the pyruvate dehydrogenase multienzyme complex. (*a*) Pyruvate:ferredoxin oxidoreductase of the archaebacteria. (*b*) Pyruvate dehydrogenase multienzyme complex of eubacteria and eukaryotes (E1, pyruvate decarboxylase; E2, dihydrolipoyl acetyltransferase; E3, dihydrolipoamide dehydrogenase). Abbreviations used: R, $-CH_3$; Fd, ferredoxin; FeS, enzyme-bound iron–sulphur cluster; TPP-H, thiamine pyrophosphate; Lip, lipoic acid; B, an amino acid base on the E3 component.

observation is in conflict with there being no detectable 6-phosphofructokinase in this archaebacterium [19]. However, it is known that in some eubacteria and eukaryotes anaerobic glycolysis relies on a pyrophosphate-dependent phosphofructokinase as opposed to an ATP-linked enzyme [22]; it would be interesting to see if the methanogens, or any other archaebacteria, contain this unusual phosphofructokinase.

The extreme halophiles are phylogenetically close to the methanogens [3,4] and they also synthesize glucose via a reverse Embden–Meyerhof pathway [10,23,24]. However, the synthesis of carbohydrates in the thermophilic archaebacteria has not been established, although in *Tp. acidophilum* many of the enzymes of the glycolytic sequence are absent [13], thus apparently ruling out a reversal of this route. A reversal of the non-phosphorylated Entner–Doudoroff pathway is energetically feasible.

Table 1. Dihydrolipoamide dehydrogenase and lipoic acid in the archaebacteria. Abbreviations used: +, indicates the presence of enzyme and/or lipoic acid; −, indicates the absence (or levels at the limits of detection) of enzyme and/or lipoic acid; nd, indicates that no data are available.

Organism	Dihydrolipoamide dehydrogenase	Lipoic acid	Reference
Halobacterium halobium	+	+	[30,33]
Haloferax volcanii	+	+	[30,32]
Natronobacterium pharaonis	+	nd	[30]
Natronobacterium gregoryi	+	nd	[30]
Natronococcus occultus	+	nd	[30]
Thermoplasma acidophilum	+	+	[67,*]
Thermococcus celer	nd	+	[32]
Methanobacterium thermoautotrophicum	+	+	[32,*]
Sulfolobus acidocaldarius	−	nd	[67]
Sulfolobus solfataricus	nd	−	[32]
Thermoproteus tenax	nd	−	[32]
Pyrodictium occultum	nd	−	[32]

* M. J. Danson, D. W. Hough, S. A. Jackman & K. J. Stevenson, unpublished work.

The conversion of pyruvate to acetyl-CoA

In all three archaebacterial phenotypes, pyruvate oxidoreductase catalyses the conversion of pyruvate into acetyl-CoA [17,25–29]. In the halophiles and thermophiles, ferredoxin serves as the electron acceptor, whereas the methanogens use the deazaflavin derivative F_{420}. The catalytic mechanism of the ferredoxin-oxidoreductase [26] is shown in Fig. 3, together with that of the pyruvate dehydrogenase multienzyme complex which serves to catalyse the oxidative decarboxylation of pyruvate to acetyl-CoA in eukaryotes and many eubacteria.

A detailed structural and mechanistic comparison between these two enzyme systems has been made previously [1]. Catalytically, both enzymes decarboxylate pyruvate in a thiamine pyrophosphate (TPP)-dependent process, the C_2 unit being handed on to CoA to form acetyl-CoA. The main mechanistic difference is to be found in this transfer process. In the archaebacterial oxidoreductase the transfer is direct from hydroxyethyl-TPP to CoA, with reducing equivalents passing to ferredoxin via protein-bound iron–sulphur centres; on the other hand, in the dehydrogenase complex a protein-bound lipoic acid residue acts as an intermediary carrier of both acetyl and reducing groups. This lipoic acid necessitates the additional presence of the enzyme dihydrolipoamide dehydrogenase in the pyruvate dehydrogenase complex.

Lipoic acid is not detectable in the purified archaebacterial pyruvate oxidoreductases [25,29]. We were therefore surprised to find the presence of dihydrolipoamide dehydrogenase in the halophilic archaebacteria [30]. Purification and characterization [31] showed it to be very similar to its complexed non-archaebacterial counterpart and, using a biological assay [32] and a g.c./m.s.

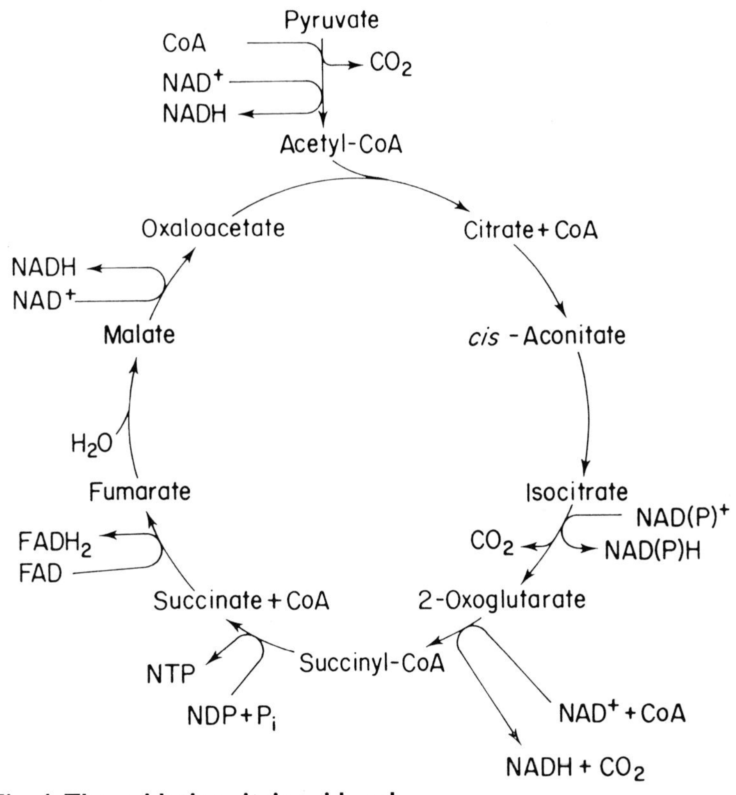

Fig. 4. The oxidative citric acid cycle.

procedure [33], its presumed substrate, lipoic acid, has also been found in these organisms. A survey of other archaebacterial genera [34] has revealed that the presence of the enzyme and lipoic acid are coincident but are found only in the methanogen–halophile branch, including *Thermoplasma* and *Thermococcus* (Table 1).

The function of this archaebacterial dihydrolipoamide dehydrogenase in the absence of its normal multienzyme complexes is unknown [34,35].

The citric acid cycle

Acetyl-CoA can be further metabolized via the citric acid cycle. As in the case of hexose catabolism, the form of this cycle can be correlated with the phenotypic characteristics of the organisms.

(*a*) The oxidative citric acid cycle. The extreme halophiles are aerobic organotrophs and possess a complete oxidative citric acid cycle [1,36,37], which will serve to generate energy and key starting materials for biosynthesis as well as permitting growth on amino acids and other nitrogenous compounds (Fig. 4). Phylogenetically related to the halophiles are *Tp. acidophilum* and *Thermoplasma volcanium*; they are facultative organotrophs [38] and when growing aerobically they are also thought to possess a complete citric acid cycle for energy generation. This may also be the case for *Sulfolobus* spp., although respirometric analyses suggest that its use may be limited [14]. A number of other archaebacteria are obligate

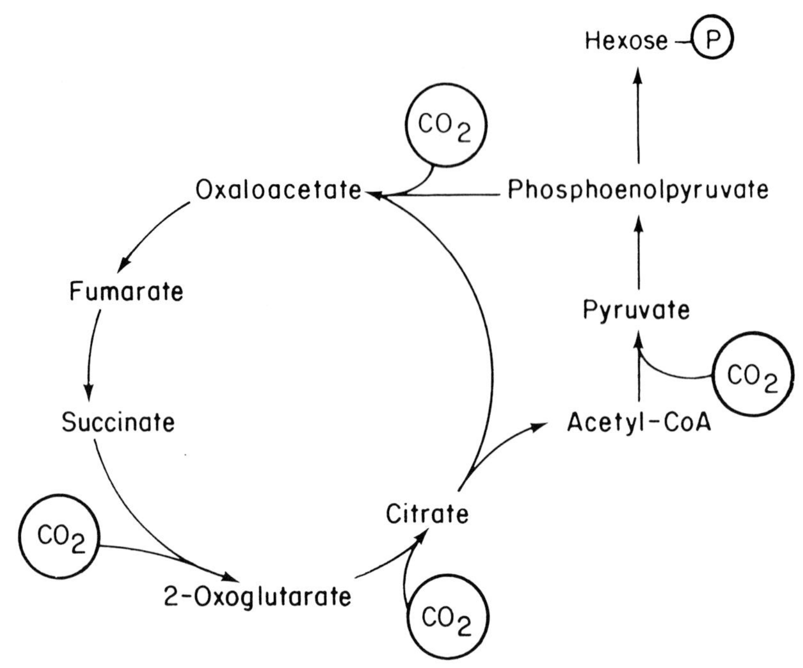

Fig. 5. The reductive citric acid cycle for autotrophic CO_2 fixation in *Thermoproteus neutrophilus*.

heterotrophs and can use amino acids for growth (e.g. *Pyrococcus*, *Thermococcus*, *Staphylothermus*, *Thermophilum* [39,40]), but there are no data yet as to whether they also possess a citric acid cycle.

(*b*) The reductive citric acid cycle. Many of the sulphur-dependent thermophilic archaebacteria can grow autotrophically, and in two genera, *Sulfolobus brierleyi* [41] and *Thermoproteus neutrophilus* [42,43], there is enzymological and metabolic evidence that CO_2 is fixed via a reductive citric acid cycle (Fig. 5). Interestingly, when *T. neutrophilus* grows on acetate, approximately 15% of 2-oxoglutarate is formed via reductive carboxylation to pyruvate and a reductive citric acid cycle, and the remainder via an oxidative cycle. Thus a horseshoe-type cycle is proposed where the two halves meet at 2-oxoglutarate. Similarly, Wood *et al.* [14] have suggested for *S. brierleyi* that the simultaneous use of acetate and CO_2 indicates that the CO_2-fixing cycle must operate simultaneously with both assimilatory and oxidative pathways.

(*c*) Partial citric acid cycles. In the methanogenic archaebacteria, two partial versions of the cycle can be observed, both fulfilling an anabolic role (Fig. 6). Thus in *M. thermoautotrophicum* [44], *Methanospirillum hungatei* [45] and *Methanococcus voltae* [46] an incomplete reductive citric acid cycle is present, leading to 2-oxoglutarate via succinate. On the other hand, *Methanosarcina barkeri* possesses an incomplete, oxidative cycle with 2-oxoglutarate synthesized via citrate [47,48]. No methanogen has yet been reported to have a complete citric acid cycle, oxidative or reductive.

(*d*) Other pathways of acetyl-CoA catabolism. Several archaebacteria may catabolize acetyl-CoA by routes other than the citric acid cycle. Mention has been

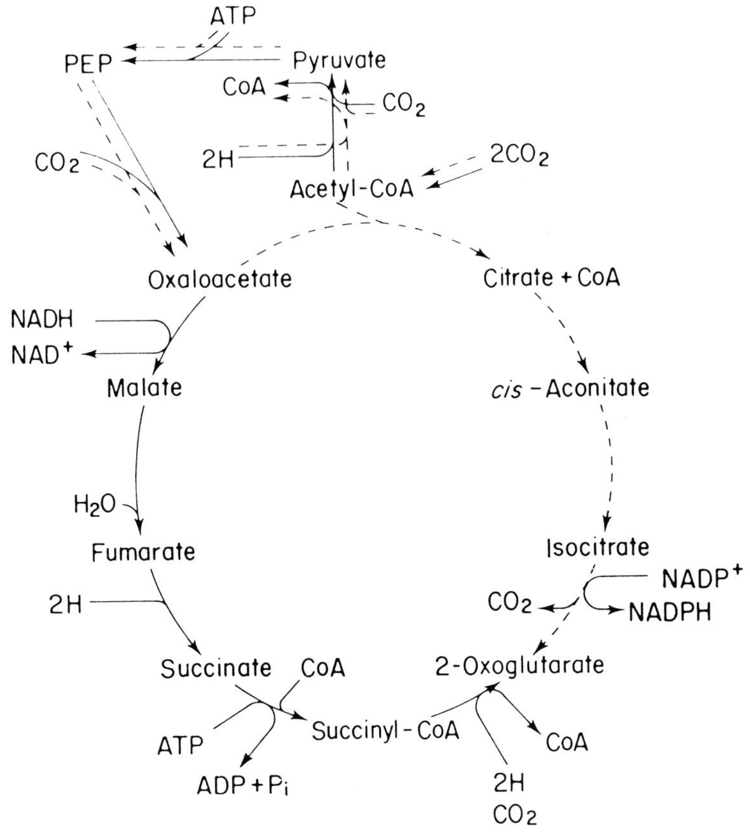

Fig. 6. The partial citric acid cycles in methanogens. The pathways shown are those proposed for *Methanobacterium thermoautotrophicum* (→) and for *Methanosarcina barkeri* (– – – – – →). Abbreviation used: PEP, phosphoenolpyruvate.

made previously that *Tp. acidophilum* and *P. furiosus* can both generate ATP by the conversion of acetyl-CoA into acetate. Also, in the extreme thermophile, *Archeoglobus fulgidus*, acetyl-CoA oxidation to CO_2 proceeds via a carbon monoxide dehydrogenase pathway [49,50]. Several of the enzymes of the citric acid cycle have been found in *A. fulgidus*, but none catalysing the oxidation of 2-oxoglutarate or succinate; thus in this organism the cycle may be operating biosynthetically with the carbon monoxide pathway serving to generate energy.

Evolution of central metabolism

With a knowledge of the routes of central metabolism in the archaebacteria, eubacteria and eukaryotes, it may now be possible to speculate on the evolutionary origin of these pathways. However, such speculations rely on a definitive phylogeny of these organisms, and in particular which evolutionary lineage is most primitive. Current thoughts [3,4] suggest that the archaebacteria are indeed members of an ancient evolutionary line, especially with respect to the eukaryotes, and that thermophily may be the ancestral phenotype.

Modifications of the Entner–Doudoroff pathway are found in both halophilic and thermophilic archaebacteria, whereas the Embden–Meyerhof sequence appears to be used in halophiles and methanogens in a gluconeogenic direction. Thus it is possible that these two pathways originated for catabolism and anabolism respectively, and that evolution of ATP-phosphofructokinase (not yet found in the archaebacteria) permitted the glycolytic sequence to operate in a degradative manner. Clearly, more archaebacterial studies are required, especially within the hyperthermophiles.

On the conversion of pyruvate into acetyl-CoA, pyruvate oxidoreductases have been found throughout the archaebacteria, and also in anaerobic eubacteria. Thus, Kerscher & Oesterhelt [29] suggest that these enzymes existed before the divergence of the three kingdoms and that the pyruvate dehydrogenase complex (found only in the respiratory eubacteria and eukaryotes) emerged after the development of oxidative phosphorylation. Similarly, we have suggested that dihydrolipoamide dehydrogenase may also be an ancient enzyme [35] as it is found in archaebacteria, eubacteria and eukaryotes. Presumably, its original function has been retained in the archaebacteria, but in the respiratory eubacteria and eukaryotes it has been sequestered to function as an integral component of the 2-oxoacid dehydrogenase complexes.

Finally, with respect to the citric acid cycle, it has been suggested [51] that the reductive conversion of oxaloacetate into succinate may have evolved in association with anaerobic hexose metabolism to regenerate NAD^+. With the evolution of the other half of the cycle, whereby 2-oxoglutarate is synthesized from acetyl-CoA and oxaloacetate via citrate, a primitive anaerobic cell may have possessed two arms of the cycle [52]. Evolution of 2-oxoglutarate oxidoreductase from the analogous pyruvate enzyme would complete the cycle which would probably act initially in a reductive mode. Thus, the reductive citric acid cycle is seen as the primitive pathway, and it is interesting that such a cycle is found in species of *Sulfolobus* and *Thermoproteus*, and incompletely in various methanogens, organisms that represent the two main archaebacterial branches.

(II) Molecular enzymology

Comparative enzymology

For the reasons outlined in the Introduction, our laboratory is currently engaged in the structural analysis of several archaebacterial enzymes of central metabolism. Three enzymes have been chosen initially, our choices being made with respect to their metabolic position and to the availability of structural data for their non-archaebacterial counterparts: (*a*) glucose dehydrogenase, the first enzyme of the modified Entner–Doudoroff pathways; (*b*) dihydrolipoamide dehydrogenase, an enzyme that we discovered in the archaebacteria while investigating the conversion of pyruvate into acetyl-CoA; and (*c*) citrate synthase, the initial enzyme of the citric acid cycle.

In each case, our strategy has been to purify the enzyme in question, to determine its catalytic and oligomeric features and an *N*-terminal amino acid sequence, and then to use this sequence to design oligonucleotide probes to effect the cloning of the desired gene. A directional cloning approach, that is, using two restriction endonucleases to digest the genomic DNA, is preferred, as is the use of a unique, long (approx. 48-mer) probe for Southern blot analysis. After cloning and DNA sequencing, expression in *Escherichia coli* has been attempted to give quantities of protein sufficient for subsequent structural analysis.

The long-term aim is the structural determination of these enzymes from a range of archaebacterial phenotypes: thermophilic, halophilic and mesophilic. As demonstrated below, the extreme stability of some of the enzymes has been crucial to their purification after expression in a mesophilic host.

Glucose dehydrogenase

Glucose dehydrogenase catalyses the reaction:

$$\text{Glucose} + \text{NAD(P)}^+ \rightarrow \text{gluconate} + \text{NAD(P)H} + \text{H}^+$$

The enzyme has been purified from the thermophile, *Tp. acidophilum*, and has been shown to be active with both NAD^+ and $NADP^+$ [53]. Thus, it is one of several dual-cofactor-specific dehydrogenases found in the thermophilic archaebacteria. The gene encoding the *Tp. acidophilum* glucose dehydrogenase has been cloned, sequenced and expressed in *E. coli* [54]. The expressed protein is fully active and we have used its inherent thermostability to effect a simple purification procedure.

The derived amino acid sequence shows less than 20% identity with the glucose dehydrogenases from the eubacteria *Bacillus subtilis* and *Bacillus megaterium*. However, using consensus dehydrogenase sequence information, a region of the *Tp. acidophilum* sequence has been identified with the characteristics of an ADP-binding fold. Conclusive identification of the nucleotide-binding site must await the determination of its three-dimensional structure, and to this end we have recently reported the crystallization of the enzyme [55].

Dihydrolipoamide dehydrogenase

Dihydrolipoamide dehydrogenase catalyses the reaction:

$$\text{Dihydrolipoamide} + \text{NAD}^+ \rightleftharpoons \text{lipoamide} + \text{NADH} + \text{H}^+$$

As discussed above, our interest in this enzyme lies in the fact that it is present in a number of archaebacteria despite the absence of the 2-oxoacid dehydrogenase complexes. We do not yet know the cellular function of the archaebacterial enzyme, and therefore it is important to gain structural information to compare with its multienzyme counterparts. The enzyme has been purified and characterized from the extremely halophilic archaebacteria [31,34]; in oligomeric structure, catalytic properties and in its active-site mechanism it is remarkably similar to the eubacterial and eukaryotic dihydrolipoamide dehydrogenases. The gene has now been cloned from *Haloferax volcanii* [56] and its sequence determined (K. J. Stevenson & N. Vettakkorumakankav, personal communication). Attempts to express an active enzyme in *E. coli* are in progress, as are similar investigations of the dihydrolipoamide dehydrogenase from *Tp. acidophilum*.

Citrate synthase

Citrate synthase catalyses the entry of carbon into the citric acid cycle:

$$\text{Oxaloacetate} + \text{acetyl-CoA} + \text{H}_2\text{O} \rightarrow \text{citrate} + \text{CoASH}$$

Table 2. Sequence identities between eukaryotic, eubacterial and archaebacterial citrate synthases. Multiple alignment of citrate synthases was carried out as described in [65].

	% Identity with citrate synthase from:							
Citrate synthase source	(1)	(2)	(3)	(4)	(5)	(6)	(7)	(8)
(1) Pig	100	60	50	23	23	21	19	20
(2) Yeast		100	48	22	22	20	19	20
(3) *Arabidopsis thaliana*			100	19	21	19	18	19
(4) *Escherichia coli*				100	62	69	57	27
(5) *Acinetobacter anitratum*					100	70	56	28
(6) *Pseudomonas aeruginosa*						100	57	27
(7) *Rickettsia prowazekii*							100	28
(8) *Thermoplasma acidophilum*								100

Of the enzymes of central metabolism, we feel that citrate synthase is one of the best for the purposes of comparative enzymology and evolutionary studies: it is present in nearly all organisms, it displays a diversity of properties that shows a remarkable correlation with the taxonomic status of the source organism, and there is a wealth of structural information on the enzyme from eubacteria and eukaryotes (reviewed in [1] and [57]).

In summary, citrate synthases from Gram-negative eubacteria are hexameric proteins and are allosterically inhibited by NADH, the primary energetic end-product of the citric acid cycle. In contrast, the enzymes from Gram-positive eubacteria and eukaryotes are all dimers; they are insensitive to NADH but are inhibited by ATP, the ultimate end product of the cycle. The subunits of all citrate synthases are approximately the same size (*ca.* 45000–50000) and we have evidence that the hexamers function as trimers of the basic dimeric unit [58]. Sequences of a number of eubacterial and eukaryotic citrate synthases are now available, as is a high-resolution structure of the pig heart enzyme [59,60].

We have demonstrated the presence of citrate synthase in all three archaebacterial phenotypes [37] and it has been shown that the enzymes from *Sulfolobus acidocaldarius*, *Tp. acidophilum* and *Hf. volcanii* are dimeric proteins with catalytic and regulatory properties similar to the dimers from Gram-positive eubacteria and eukaryotes [61–64].

The gene encoding citrate synthase from *Tp. acidophilum* has been cloned and sequenced and the derived amino acid sequence compared by multiple alignment with other citrate synthases [65]. The sequence identities between the three groups, archaebacteria, eubacteria and eukaryotes, are $< 30\%$, consistent with the three separate domains (Table 2) [65]. However, the majority of residues implicated in the catalytic action of the enzyme have been conserved across all organisms. Interestingly, the archaebacterial citrate synthase lacks approximately the first 50 amino acids (the two *N*-terminal helices) of the eubacterial and eukaryotic enzymes. Also, there are 13 positions within the sequence that are identical in all eubacterial and eukaryotic citrate synthases but differ in the *Tp. acidophilum* enzyme. Of these 13, seven lie close to the subunit interface in the pig enzyme, and it is tempting to speculate that they might play a role in the thermostability of the archaebacterial protein. Currently, we are investigating this possibility using site-directed mutagenesis.

The *Tp. acidophilum* citrate synthase gene has been expressed in *E. coli* to give the active enzyme and the thermostability used to effect a simple purification procedure [66]. Detailed structural studies are now possible.

We have also purified the citrate synthase from *Hf. volcanii* and determined an *N*-terminal sequence [64]. It aligns with the *N*-terminus of the *Tp. acidophilum* enzyme, and is therefore also approximately 50 residues shorter than the non-archaebacterial proteins. The cloning and sequencing of the halophilic citrate synthase gene are now in progress.

(III) Conclusions

The establishment of metabolic pathways for the production of energy and for biosynthesis would have been the basis for survival and successful competition during early life on earth. Given that we now have a reliable and quantitative phylogenetic tree of all living organisms, and the strong indication that the archaebacterial lineage is indeed ancient, an investigation of archaebacterial central metabolism may give us vital information on the nature of the earliest organisms and their subsequent evolution. Furthermore, studies on archaebacterial metabolism have given new directions to the field of comparative enzymology and to our investigations of the structural basis of protein stability. The detailed study of archaebacterial enzymes therefore remains a priority.

Financial support from the S.E.R.C., NATO and ICI Biological Products, Billingham, U.K. is gratefully acknowledged. Special thanks are also due to colleagues Dr K. J. Stevenson and N. Vettakkorumakankav (University of Calgary, Canada), Dr M. J. Bonete (University of Alicante, Spain), Dr D. Byrom (ICI Biological Products, Billingham, U.K.) and Dr J. R. Bright, Dr N. Budgen, K. D. James, R. J. M. Russell, Dr L. D. Ruttersmith, Dr K. J. Sutherland and Dr P. Towner (University of Bath, U.K.).

References

1. Danson, M. J. (1988) Adv. Microbiol. Physiol. **29**, 165–231
2. Danson, M. J. (1989) Can. J. Microbiol. **35**, 58–64
3. Woese, C. R., Kandler, O. & Wheelis, M. L. (1990) Proc. Natl. Acad. Sci. U.S.A. **87**, 4576–4579
4. Woese, C. R. (1987) Microbiol. Rev. **51**, 221–271
5. Cooper, R. A. (1986) in Carbohydrate Metabolism in Cultured Cells (Morgen, M. J., ed.), pp. 461–491, Plenum Press, New York and London
6. Perham, R. N., Packman, L. C. & Radford, S. E. (1987) Biochem. Soc. Symp. **54**, 67–81
7. Morris, J. G. (1985) in Comprehensive Biotechnology (Moo-Young, M., ed.), pp. 357–378, Pergamon Press, Oxford

8. Tomlinson, G. A., Koch, T. K. & Hochstein, L. I. (1974) Can. J. Microbiol. **20**, 1085–1091
9. Hochstein, L. I. (1988) in Halophilic Bacteria (Rodriguez-Valera, F., ed.), vol. 2, pp. 67–83, CRC Press Inc., Boca Raton, U.S.A
10. Rawal, N., Kelkar, S. M. & Altekar, W. (1988) Ind. J. Biochem. Biophys. **25**, 674–686
11. Severina, L. O. & Pimenov, N. V. (1988) Mikrobiologiya **57**, 907–911
12. De Rosa, M., Gambocorta, A., Nicolaus, B., Giardina, P., Poerio, E. & Buonocore, V. (1984) Biochem. J. **224**, 407–414
13. Budgen, N. & Danson, M. J. (1986) FEBS Lett. **196**, 207–210
14. Wood, A. P., Kelly, D. P. & Norris, P. R. (1987) Arch. Microbiol. **146**, 382–389
15. Muckund, S. & Adams, M. W. W. (1991) J. Biol. Chem. **266**, 14208–14216
16. Budgen, N. (1988) Ph.D. Thesis, University of Bath, Bath, U.K.
17. Schafer, T. & Schönheit, P. (1991) Arch. Microbiol. **155**, 366–377
18. Jansen, K., Stupperich, E. & Fuchs, G. (1982) Arch. Microbiol. **132**, 355–364
19. Fuchs, G., Winter, H., Steiner, I. & Stupperich, E. (1983) Arch. Microbiol. **136**, 160–162
20. Evans, J. N. S., Tolman, C. J., Kanodia, S. & Roberts, M. F. (1985) Biochemistry **24**, 5693–5698
21. Evans, J. N. S., Raleigh, D. P., Tolman, C. J. & Roberts, M. F. (1986) J. Biol. Chem. **261**, 16323–16331
22. Mertens, E. (1991) FEBS Lett. **285**, 1–5
23. D'Souza, S. A. & Altekar, W. (1983) Ind. J. Biochem. Biophys. **20**, 29–32
24. Altekar, W. & Rajagopalan, R. (1990) Arch. Microbiol. **153**, 169–174
25. Kerscher, L. & Oesterhelt, D. (1981) Eur. J. Biochem. **116**, 587–594
26. Kerscher, L. & Oesterhelt, D. (1981) Eur. J. Biochem. **116**, 595–600
27. Kerscher, L., Nowitzki, S. & Oesterhelt, D. (1982) Eur. J. Biochem. **128**, 223–230
28. Zeikus, J. G., Fuchs, G., Kenealy, W. & Thauer, R. K. (1977) J. Bacteriol. **132**, 604–613
29. Kerscher, L. & Oesterhelt, D. (1982) Trends Biochem. Sci. **7**, 371–374
30. Danson, M. J., Eisenthal, R., Hall, S., Kessell, S. R. & Williams, D. L. (1984) Biochem. J. **218**, 811–818
31. Danson, M. J., McQuattie, A. & Stevenson, K. J. (1986) Biochemistry **25**, 3880–3884
32. Noll, K. M. & Barber, T. S. (1988) J. Bacteriol. **170**, 4315–4321
33. Pratt, K. J., Carles, C., Carne, T. J., Danson, M. J. & Stevenson, K. J. (1989) Biochem. J. **258**, 749–754
34. Danson, M. J., Hough, D. W., Vettakkorumakankav, N. & Stevenson, K. J. (1991) in General and Applied Aspects of Halophilic Micro-organisms, NATO ASI Ser. A (Rodriguez-Valera, F., ed.), vol. 201, pp. 121–128, Plenum Publishing Corp., New York
35. Danson, M. J. (1988) Biochem. Soc. Trans. **16**, 87–89
36. Aitken, D. M. & Brown, A. D. (1969) Biochim. Biophys. Acta **177**, 351–354
37. Danson, M. J., Black, S. C., Woodland, D. L. & Wood, P. A. (1985) FEBS Lett. **179**, 120–124
38. Segerer, A., Langworthy, T. A. & Stetter, K. O. (1988) Syst. Appl. Microbiol. **10**, 161–171
39. Stetter, K. O. (1986) in Thermophiles, General and Applied Microbiology (Brock, T. D., ed.), pp. 39–74, John Wiley, New York
40. Kelly, R. M. & Demming, J. W. (1988) Biotech. Prog. **4**, 47–62
41. Kandler, O. & Stetter, K. O. (1981) Zentrabl. Bakteriol. Hyg. Abt. 1 Orig. **C2**, 111–121
42. Schafer, S., Barkowski, C. & Fuchs, G. (1986) Arch. Microbiol. **146**, 301–308

43. Schafer, S., Paalme, T., Vilu, R. & Fuchs, G. (1989) Arch. Microbiol. **186**, 695–700
44. Fuchs, G. & Stupperich, E. (1982) in Archaebacteria (Kandler, O., ed.), pp. 277–288, Gustav Fischer, Stuttgart
45. Ekiel, I., Smith, I. C. P. & Sprott, G. D. (1983) J. Bacteriol. **156**, 316–326
46. Ekiel, I., Jarrell, K. F. & Sprott, G. D. (1985) Eur. J. Biochem. **149**, 437–444
47. Daniels, L. & Zeikus, J. G. (1978) J. Bacteriol. **136**, 75–84
48. Weimer, P. J. & Zeikus, J. G. (1979) J. Bacteriol. **137**, 332–339
49. Thauer, R. K., Möller-Zinkhan, D. & Spormann, A. M. (1989) Annu. Rev. Microbiol. **43**, 43–67
50. Möller-Zinkhan, D. & Thauer, R. K. (1990) Arch. Microbiol. **153**, 215–218
51. Gest, H. (1987) Biochem. Soc. Symp. **54**, 3–16
52. Weitzman, P. D. J. (1985) in Evolution of Prokaryotes (Schleifer, K. H. & Stackenbrandt, E., eds.), pp. 253–275, Academic Press, London
53. Smith, L. D., Budgen, N., Bungard, S. J., Danson, M. J. & Hough, D. W. (1989) Biochem. J. **261**, 973–977
54. Bright, J. R., Byrom, D., Danson, M. J., Hough, D. W. & Towner, P. (1991) Eur. J. Biochem. in the press
55. Bright, J. R., Mackness, R., Danson, M. J., Hough, D. W., Taylor, G. L., Towner, P. & Byrom, D. (1991) J. Mol. Biol. **222**, 143–144
56. Vettakkorumakankav, N., Stevenson, K. J. & Danson, M. J. (1991) in Flavins and Flavoproteins (Curti, B., Ronchi, S. & Zanetti, G., eds.), in the press, Walter de Gruyter & Co., Berlin
57. Weitzman, P. D. J. & Danson, M. J. (1976) Curr. Top. Cell. Regul. **10**, 161–204
58. Else, A. J., Danson, M. J. & Weitzman, P. D. J. (1988) Biochem. J. **251**, 803–807
59. Remington, S., Wiegand, G. & Huber, R. (1982) J. Mol. Biol. **158**, 111–152
60. Wiegand, G., Remington, S., Diesenhofer, J. & Huber, R. (1984) J. Mol. Biol. **174**, 205–219
61. Smith, L. D., Stevenson, K. J., Hough, D. W. & Danson, M. J. (1987) FEBS Lett. **225**, 277–281
62. Lohlein-Werhahn, G., Goepfert, P. & Eggerer, H. (1988) Biol. Chem. Hoppe-Seyler **369**, 109–113
63. Grössebuter, W. & Görisch, H. (1985) Syst. Appl. Microbiol. **6**, 119–124
64. James, K. D., Bonete, M. J., Byrom, D., Danson, M. J. & Hough, D. W. (1991) Biochem. Soc. Trans. **20**, 12S
65. Sutherland, K. J., Henneke, C. M., Towner, P., Hough, D. W. & Danson, M. J. (1990) Eur. J. Biochem. **194**, 839–844
66. Sutherland, K. J., Danson, M. J., Hough, D. W. & Towner, P. (1991) FEBS Lett. **282**, 132–134
67. Smith, L. D., Bungard, S. J., Danson, M. J. & Hough, D. W. (1987) Biochem. Soc. Trans. **15**, 1097

Biochem. Soc. Symp. **58**, 23–39
Printed in Great Britain

Bioenergetics and autotrophic carbon metabolism of chemolithotrophic archaebacteria

Georg Fuchs, Axel Ecker and Gerhard Strauss

Abteilung Angewandte Mikrobiologie, University of Ulm, W-7900 Ulm, F.R.G.

Introduction

In the past two decades the discovery of most of the archaebacterial species known today has broadened our view of the microbial world. The knowledge of their distinct phylogenetic status, their interesting structural features and metabolic capabilities, and their adaptation to mostly harsh environmental conditions not only had a great scientific impact but also some emerging applications are based largely on these properties. It is generally believed that the separation of life forms into at least three main lines took place at a time when the average temperature on earth was much higher than today and molecular oxygen was essentially absent. It appears that archaebacteria more than eubacteria have conserved what are considered primitive physiological and metabolic traits, especially life at high temperatures and life without oxygen.

In fact, the majority of the described archaebacterial species are anaerobes. A considerable number of them are characterized by a chemolithoautotrophic mode of growth. Their energy metabolism, as far as it is known, is based mainly on three types of anaerobic respiration: (1) the reduction of CO_2 or C_1 compounds to methane, or the disproportionation of acetate to CO_2 and methane; (2) the reduction of sulphate or other oxosulphur compounds to hydrogen sulphide; and (3) the reduction of elemental sulphur to hydrogen sulphide. Interestingly, the capability to use methane formation for energy conservation is restricted to the methanogenic branch of the archaebacteria; it has not been found in any other group of organisms; *vice versa*, CO_2 reduction to acetate, which has many things in common with methanogenesis from CO_2 or acetate, is used in energy metabolism in eubacteria only [1]. (A similarly singular metabolic trait of archaebacteria is the phototrophic growth of halophilic archaebacteria.) In contrast, sulphur respiration is also found in eubacteria.

Most of the chemolithotrophic archaebacteria are able to synthesize all cellular compounds from CO_2 via non-Calvin-type mechanisms. Two different autotrophic carbon fixation pathways are found in archaebacteria which occur also in eubacteria. There may be even more ways of autotrophic life in archaebacteria as well as in eubacteria.

This contribution deals with those archaebacteria which grow in the complete absence of molecular oxygen, with inorganic electron acceptors and which use H_2 and CO_2 as sole reductant and carbon source, respectively. They obtain energy by anaerobic respiration. These bacteria are interesting, first, because of the seemingly striking simplicity of their energy and carbon metabolism and, secondly, because their metabolic capabilities have allowed them to survive under conditions which may be not too different from the early physical and chemical environment of some three billion years ago. The energetics of the anaerobic respirations used by archaebacteria will be surveyed first. Then a review of the autotrophic pathways used by archaebacteria and evidence for a new autotrophic pathway in the phototrophic green eubacterium *Chloroflexus aurantiacus* will be presented. Only the most recent literature will be considered, since this research area has been reviewed extensively [1–18].

Energy conservation process in anaerobic chemolithotrophic archaebacteria

Energy conservation in the chemolithotrophic anaerobic archaebacteria is based on different types of anaerobic respirations in which the oxidation of hydrogen is coupled to the reduction of inorganic compounds (Fig. 1). The difference in the redox potential between H_2/H^+ and the inorganic redox couple is used to drive ATP synthesis by a chemiosmotic mechanism. The three processes considered, CO_2 reduction, sulphate reduction and sulphur reduction, have in common that they are not exergonic enough to allow the synthesis of one ATP/reaction. This is owing to the following reasons. (1) The inorganic electron acceptors CO_2, sulphur, and sulphate are weak oxidants. In addition the hydrogen partial pressure found in most parts of anaerobic environments is only one thousandth of standard conditions. The free-energy changes under non-standard conditions associated with these H_2-dependent reductive reactions are therefore far less than indicated in Fig. 1, namely in the order of only -30 to -50 kJ. The synthesis of one mole of ATP under standard conditions is endergonic by $+32$ kJ, under cellular conditions by $+50$ kJ. An energy-yielding process which operates at an overall thermodynamic efficiency of 60% therefore must be associated with a free-energy change of approx. -80 kJ at cellular activities of the reactants and products to drive the synthesis of one mole of ATP/mole of reactant. If one assumes a H^+/P ratio of 3:1 for H^+-ATP synthase, then the minimum requirement of an energy-yielding process would be the translocation of one positive charge across the membrane; this would be equivalent to one third of an ATP. (2) The redox potential of the redox couples H_2/H^+, CO_2/CH_4, SO_4^{2-}/H_2S, and S/H_2S are in the range between -400 mV and -200 mV. This implies that the electron transport chains involved cannot be conventional ones. (3) One major energetic problem is the circumstance that the

Chemolithotrophy in anaerobic archaebacteria

1. $CO_2 + 4\ H_2 \longrightarrow CH_4 + 2H_2O$
   $\Delta G_o'$ −131 kJ/mol
   Methanogens
2. $SO_4^{2-} + H^+ + 4H_2 \longrightarrow HS^- + 4H_2O$
   $\Delta G_o'$ −152 kJ/mol
   Archaeoglobus
3. $S + H_2 \longrightarrow HS^- + H^+$
   $\Delta G_o'$ −28 kJ/mol
   Thermoproteus,
   Acidianus,
   Desulfurolobus
   Pyrodictium,
   Pyrobaculum
   (autotrophic species)
4. Uncertain or little studied
   NO_3^- reduction
   Fumarate reduction
   Fe(III) reduction
   Mo(VI) reduction

Fig. 1. Chemolithotrophic growth of anaerobic archaebacteria and energetics of the energy-yielding reactions under standard conditions.

eight-electron reduction of CO_2 to CH_4 and of sulphate to H_2S first involves endergonic partial reactions, i.e. CO_2 reduction to the formaldehyde level and sulphate reduction to the sulphite level. This means that initially energy has to be spent before energy can be obtained in the following exergonic partial reactions. The energy barrier is overcome differently: by a reversed electron transport in the reduction of CO_2 to the formyl level which is catalysed by a membrane-bound oxidoreductase, or by activating the substrate sulphate with ATP, forming a sulphuric acid–phosphoric acid anhydride (APS) which then can be reduced by H_2. The sulphate-activating and the APS-reducing enzymes are soluble.

Besides CO_2 or C_1 compounds, oxosulphur compounds and elemental sulphur other inorganic electron acceptors may turn out to be used as terminal electron acceptors by some archaebacteria, but so far there is no strong evidence for other types of anaerobic respiration.

Sulphur respiration

A number of anaerobic or facultatively anaerobic sulphur-reducing archaebacteria of different genera have been described, but so far no enzymic or bioenergetic studies have been performed with respect to sulphur reduction. Therefore, the answer to the question as to how these organisms obtain energy must be speculative; it depends entirely on the limited knowledge obtained from the study

Sulphur reduction

$$S + H_2 \longrightarrow HS^- + H^+ \quad \Delta G_o' \ -28 \text{ kJ/mol}$$

$$S_8 + 2\,HS^- \rightleftharpoons 2\,S_5^{2-} + 2H^+ \quad k_{eq} = 4 \times 10^{-9}\ \text{M}$$

Sulphur → Pentasulphide

$$S_n^{2-} + 2e^- + 2H^+ \xrightarrow{H_2 \text{ or } HCOOH} S_{n-1}^{2-} + HS^- + H^+$$

- S_8 is virtually insoluble (5 μg/l)
- The pentasulphide concentration strongly depends on the pH and [HS^-]
- At pH 8.5 (25 °C) and at a sulphide concentration of 3.8 mM, 6.4 mmole elemental sulphur (in the form of S_8) can be dissolved as polysulphides/litre

Fig. 2. Basic reactions involved in sulphur reduction to H_2S as studied in *Wolinella succinogenes*.

of the Gram-negative sulphur-reducing eubacterium *Wolinella succinogenes* (Figs. 2 and 3) [13,19,19a and literature cited within]. In this organism, sulphur reduction with H_2 to H_2S is coupled to the formation of a proton motive force. One main problem of sulphur reduction is the poor solubility of elemental sulphur, which exists as S_8 and is dissolved only at a concentration of 5 μg/l. S_8 therefore has to be converted into a polysulphide, mostly pentasulphide, by reacting with two hydrogen sulphide anions. This HS^--dependent reaction strongly depends on the ambient H_2S concentration and on the pH (H_2S, $pK = 7$). It has been estimated that a pH 8.5 (25 °C) and at a HS^- concentration of 3.8 mM, 6.4 mmol elemental sulphur (or 0.8 mmol S_8) can be dissolved as polysulphides/litre. The polysulphide is the actual substrate of sulphur reduction. This process is catalysed by a complex system consisting of at least three proteins the genes of which are organized on an operon, the polysulphide reductase operon [19a]. The enzyme reducing the (S_n^{2-})polysulphide to H_2S and (S_{n-1}^{2-})polysulphide, polysulphide reductase, is an 81 kDa protein which is membrane associated; its catalytic site is oriented towards the periplasmic side of the membrane. The gene contains a consensus sequence for the binding site of molybdenum cofactor, suggesting that polysulphide reductase might be a novel molybdoenzyme. (There is also a soluble 15 kDa sulphide dehydrogenase of unknown function in this organism.) Two other open reading frames were found, one for a 21 kDa hydrophilic protein with putative iron–sulphur centre(s) and another one for a hydrophobic 34 kDa protein. Hydrogenase and formate dehydrogenase are membrane associated with the catalytic centres facing towards the periplasmic space. Preliminary experiments suggest that the reduction of sulphur (i.e. polysulphide) yields half the amount of ATP which is obtained in fumarate reduction

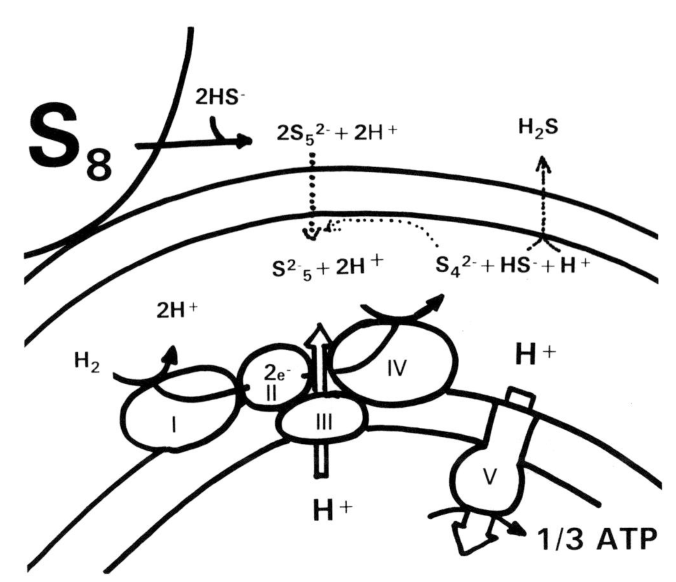

Fig. 3. Hypothetical scheme of sulphur solubilization to polysulphide, polysulphide reduction to H_2S by H_2, and coupling of this reaction to the synthesis of ATP in Gram-negative eubacteria. This scheme should be regarded only as working hypothesis for future work with archaebacterial sulphur respiration. Abbreviations: (I) hydrogenase, (II) electron carrying protein, (III) proton translocating protein, (IV) polysulphide reductase, (V) H^+-ATP synthase.

with hydrogen. Fumarate reduction probably yields two thirds of an ATP by translocation of two H^+/reaction. Therefore polysulphide reduction should be coupled with the translocation of one proton/reaction or the formation of a third of an ATP. Since both hydrogenase and polysulphide reductase have the same outside orientation, one has to postulate that the two-electron transfer is coupled to the translocation of one proton. It should be stressed that in most heterotrophic archaebacteria it has not been shown that sulphur reduction is coupled to electron-transport phosphorylation, and that the evidence in the autotrophic bacteria relies entirely on the demonstration of significant growth on sulphur, H_2 and CO_2 alone.

It is not clear how sulphur is solubilized at strongly acidic pH values by the proposed mechanism. In contrast to *W. succinogenes*, which can grow on sulphur only at neutral or slightly alkaline pH, some sulphur-reducing archaebacteria grow at a very acidic pH of 2. *Wolinella* does not seem to be attached to the sulphur globules whereas the archaebacteria studied are tightly bound to sulphur at the beginning of growth when the cell numbers are low. When the cell numbers increase the yellow elemental sulphur layer which swims on top of the culture turns greenish and finally sinks. This process can be initiated by adding cell-free supernatant of a dense culture but not by adding H_2S; this phenomenon is not understood. The growth yield of the sulphur-reducing archaebacterium *Thermoproteus neutrophilus* in batch culture is lower than 1 g of cell matter formed/mole of sulphur reduced [20] which suggests that there was metabolic uncoupling.

The following aspects, among others, need to be addressed in sulphur-reducing archaebacteria to better understand one of the most fascinating, and possibly most primitive, ways of energy conservation in living beings: How many enzymes are involved in sulphur reduction and are they functionally related to the eubacterial proteins? Are their genes organized in an operon? Which proteins are membrane bound and how are their active sites oriented? How is sulphur solubilized and which is the active species of sulphur reduced? What is the electron mediator(s) between hydrogenase and polysulphide reductase? How is electron transport coupled to the translocation of H^+ and what is the stoichiometry of that reaction? What is the benefit of binding the cells to the sulphur surface and what causes binding? Is there an active transport of polysulphide in the case where enzymes are membrane associated but face towards the cytoplasm? Are there similarities or even homologies between the eubacterial and the archaebacterial sulphur-reducing systems? What are the properties of H^+-ATP synthase?

Sulphate respiration

More is known about the reduction of sulphate by H_2 in the sulphate-reducing archaebacteria which belong to the genus *Archaeoglobus* [16,21]. Here also, the energetic coupling has only been studied in sulphate-reducing eubacteria. It remains to be shown to what extent the processes in the archaebacteria and the eubacteria are similar. The first comparison of genes suggests similarity between these groups, and the basic enzymic steps seem to be identical. There are differences, however, in that *Archaeoglobus* uses a set of novel coenzymes or modified coenzymes in hydrogen and carbon metabolism which are typical for methanogenic bacteria. These coenzymes include deazaflavin as an electron carrier and methanofuran and methanopterin as carriers for C_1 groups. This variation is explained by the fact that the sulphate-reducing and the methanogenic archaebacteria have common roots.

Sulphate (SO_4^{2-}) uptake is best studied in *Desulfovibrio*. It can be transported in an electrogenic symport with three positive charges (H^+ or, in marine organisms, Na^+) (Fig. 4). Two H^+ are consumed by reacting with the product S^{2-} to give volatile H_2S which freely passes through the membrane. Sulphate transport is therefore associated with the net uptake of 1 H^+ which is equivalent to one-third of an ATP. The next energy-consuming step is the activation of sulphate by soluble ATP sulphurylase (EC 2.7.7.4). In this reaction pyrophosphate is formed, which appears not to be used to drive phosphorylations but is hydrolysed. Sulphate activation is therefore equivalent to two ATP. Adenylylsulphate (APS) is reduced to sulphite by soluble APS reductase (EC 1.8.99.2), an FAD enzyme, and sulphite is reduced to sulphide by a possibly membrane-associated sulphite reductase which contains sirohaem [19b,19c]. The prosthetic groups of these two enzymes are those found in eubacteria. A general scheme of the energetics of sulphate reduction is given in Fig. 5.

Since sulphate transport and activation require a minimum of two ATP (assuming that no use is made of pyrophosphate), sulphite reduction must be associated with the synthesis of more than that amount of ATP. A model of how the synthesis of approx. three ATP molecules may be accomplished in the Gram-negative

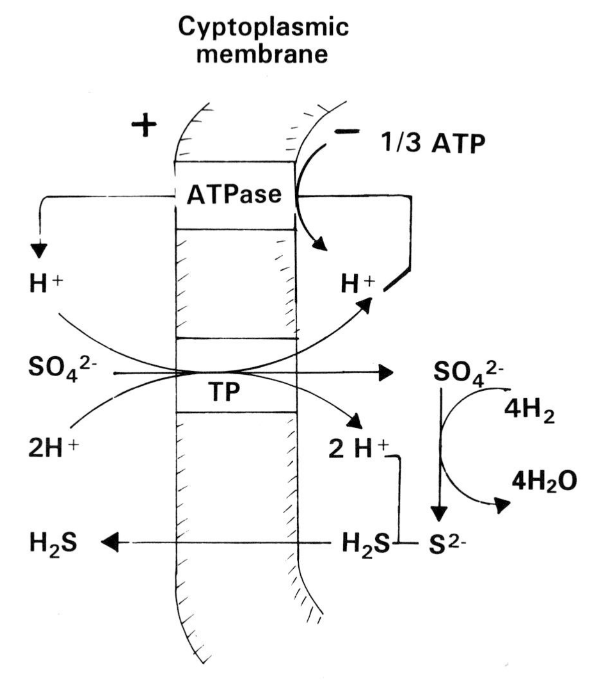

Fig. 4. Sulphate transport in Gram-negative eubacteria. The transport in archaebacteria has not been studied [21]. Abbreviation used: TP, transport protein.

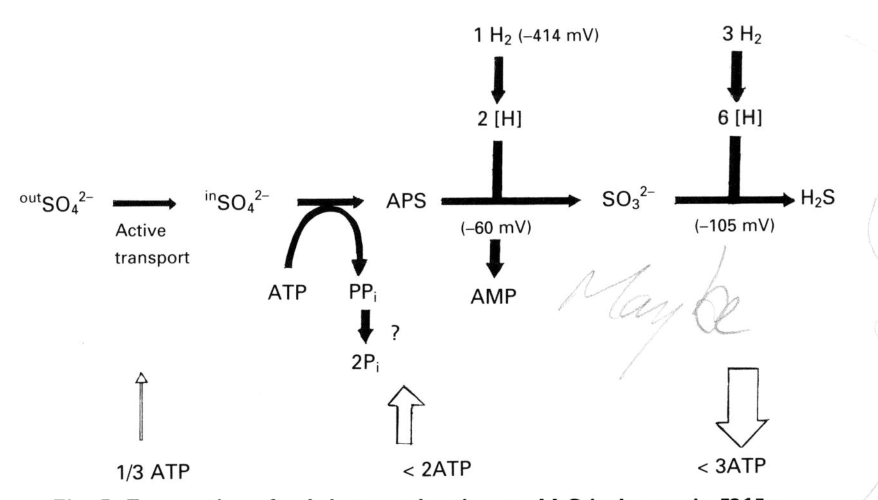

Fig. 5. Energetics of sulphate reduction to H_2S in bacteria [21].

eubacterial genus *Desulfovibrio* is given in Fig. 6. A proton motive force is created by the fact that the H_2-oxidizing, proton-releasing site of the membrane-bound hydrogenase is oriented to the outside and only electrons are transduced over the membrane. Sulphite reductase has the opposite orientation and consumes six

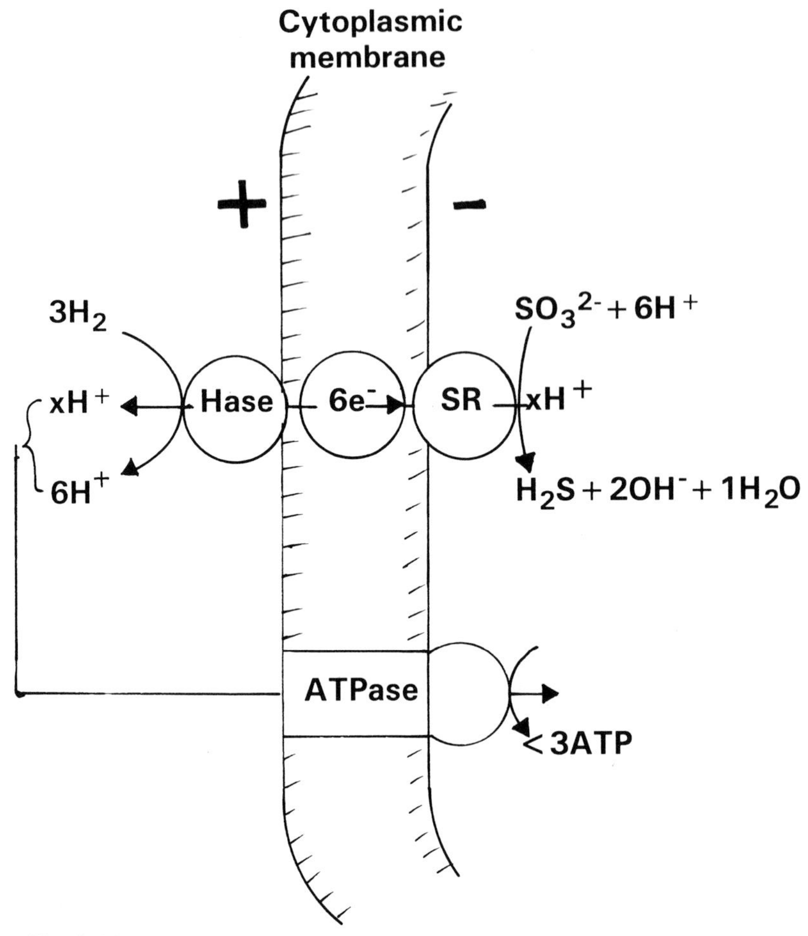

Fig. 6. Model for energy conservation linked to sulphite reduction to H_2S [21]. Abbreviations used: Hase, hydrogenase; SR, sulphite reductase.

electrons plus six H^+ from the cytoplasmic site. This process is thought to be linked to the translocation of additional H^+ otherwise the ATP yield would not be sufficient even for sulphate transport and activation alone. This hypothetical scheme needs to be tested in Archaebacteria.

CO_2 or C_1 respiration

Methanogenic bacteria obtain energy by reducing CO_2 or C_1 compounds such as CH_3OH or CH_3NH_2 with H_2 or formate to CH_4. Acetate is disproportionated; this process requires ATP-dependent activation of acetate to acetyl-CoA and cleavage to an enzyme-bound metal carbonyl (CO) group and a tetrahydromethanopterin-bound methyl group. The oxidation of $-CO$ to CO_2 is coupled to the reduction of $-CH_3$

Methanogenesis from CO_2			E_0' (mV)	$\Delta G_0'$ (kJ/mol) $2[H] = H_2$
(1)	CO_2 + MFR + 2[H]	→ CHO – MFR + H_2O	–496	+16
(2)	CHO – MFR + H_4MPT	→ CCHO – H_4MPT + MFR		–4
(3)	CHO – H_4MPT + H^+	→ CCH ≡ H_4MPT^++ H_2O		–5
(4)	CH ≡ H_4MPT^+ + 2[H]	→ CCH_2 = H_4MPT + H^+–	386	–5
(5)	CH_2 = H_4MPT + 2[H]	→ CH_3 – H_4MPT	–323	–18
(6)	CH_3 – H_4MPT + H-S-CoM	→ CH_3-S-CoM + H_4MPT		–30
(7)	CH_3-S-CoM + H-S-HTP	→ CH_4 + CoM-S-S-HTP		–45
(8)	CoM-S-S-HTP + 2[H]	→ H-S-CoM + H-S-HTP	–210	–40
(9)	CO2 + 8[H]	→ CH_4 + $2H_2O$	–244	–131

Methanogenesis from acetate			E_0' (mV)	$\Delta G_0'$ (kJ/mol) $2[H] = H_2$
(10)	CH_3-CO– + ATP^{4-}	→ CH_3-CO-O-PO_4^{2-} + ADP^{3-}		+13
(11)	CH_3-CO-O-PO_3^{2-} + H-S-COA	→ CH_3-CO-S-COA + HPO_4^{2-}		–9
(12)	CH_3-Co-O-PO_3^{2-} + H_4MPT	→ CH_3-MPT + CO + H-S-CoA		+63 (+37)
(13)	CO + H_2O	→ CO_2 + 2[H}	–524	–20
(14)	CH_3-MPT + H-S-CoM	→ CH_3 S-CoM + H_4MPT		–30
(15)	CH_3-S-CoM + H-S-HTP	→ CH_4 + CoM-S-S-HTP		–45 } (–58)
(16)	CoM-S-S-HTP + 2[H]	→ H-S-CoM + H-S-HTP	–110	–40 }
(17)	CH_3COO^- + H^+	→ CH_4 + CO_2		–36

Fig. 7. Reactions involved in methanogenesis from CO_2 and from acetate. The free-energy changes in parentheses were estimated from analogous reactions in acetogenic bacteria; they represent the lower limits [6,12,15]. Abbreviations used: MFR, methanofuran; H_4MPT, tetrahydromethanopterin; CoM-S-H, coenzyme M (mercaptoethanesulphonate); HTP-S-H, mercaptoheptanoylthreoninephosphate.

to CH_4. The energy metabolism of methanogens has much in common with the total synthesis of acetate in several eubacteria (Fig. 7). Only the last step, the reduction of the methyl group to CH_4, is unique (for reviews see [2,3,5,9–12,14,15,17,18]).

To understand methanogenesis from CO_2 one has to take into account: (i) CO_2 is reduced in four sequential two-electron reduction steps, however, the C_1 intermediates formate, formaldehyde, and methanol do not occur in a free form but are coenzyme-bound. Several new coenzymes and prosthetic groups were discovered [5,12,15,17,18] (for biochemical aspects of methanogenesis see the contribution by Wolfe, pp. 41–49, in this book). (ii) A proton motive force builds up, but in addition methanogenesis requires sodium ions and a sodium motive force [10,15,22]. (iii) The

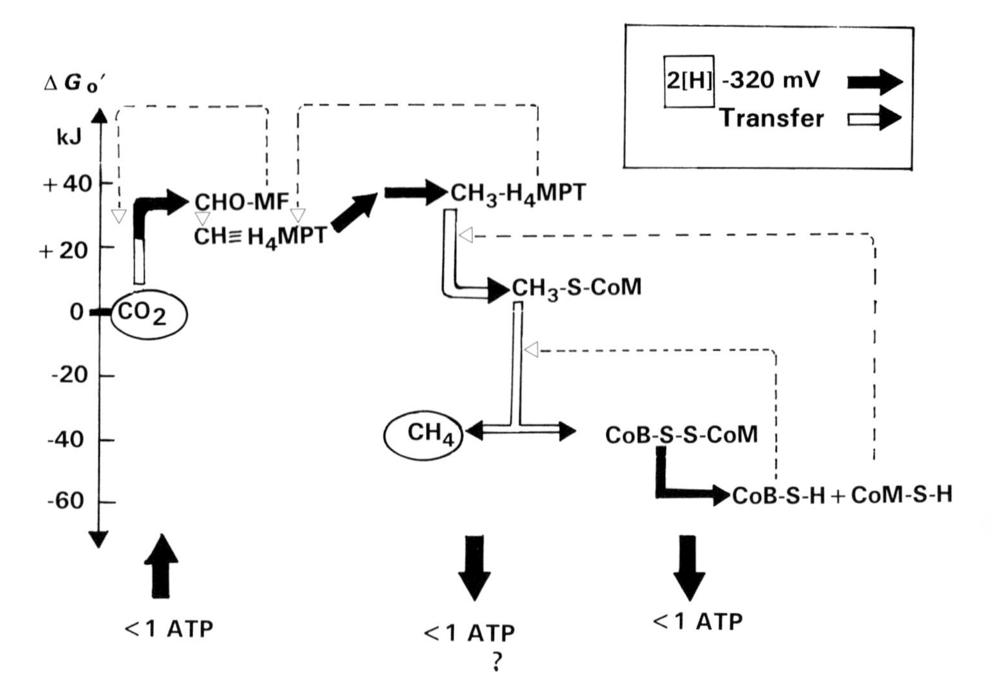

Fig. 8. Scheme of the energetics of the reactions involved in CO_2 reduction to methane. The redox reactions are standardized based on the assumption that all reductants in the four two-electron steps have a half side potential E_0' of -320 mV. This E_0' value corresponds to the average redox potential of H_2/H^+ under natural conditions (pH 7, 10^{-3}atm H_2). The Gibb's free-energy values of several of the reactions are not certain and were estimated. Black arrows represent redox reactions, white arrows represent C_1-transfer reactions. Abbreviations used: MF, methanofuran; H_4MPT, tetrahydromethanopterin; CoM-S-H, coenzyme M (mercaptoethanesulphonate); CoB-S-H, mercaptoheptanoylthreonine-phosphate (formerly designated component B, now abbreviated H-S-HTP); CHO–, formyl; CH≡, methenyl; CH_3–, methyl; ↑, ATP input; ↓, ATP output. Note that CH_3-S-CoM reduction to CH_4 is an internal redox reaction in which CoB-S-H acts as reductant for methane release and thereby forms a mixed disulphide with CoM.

endergonic reduction of CO_2 to the coenzyme-bound formyl group is driven by a reversed electron flow, coupled to the exergonic reduction of the coenzyme-bound C_1 unit to CH_4. There are probably three different energy levels in the course of CO_2 reduction to methane [15] (Fig. 8).

(1) CO_2 is bound to methanofuran and subsequently reduced by the same enzyme to the formyl level; this endergonic reduction is probably driven by a sodium motive force ($Na_o/Na_i > 10$) and is associated with the flow of two to three Na^+ into the cell. ATP as a driving force has been excluded. The corresponding enzyme, formylmethanofuran dehydrogenase, is membrane associated, and this is also likely to be true for the electron donor. The reduction of the formyl group to the

Methyltransferase

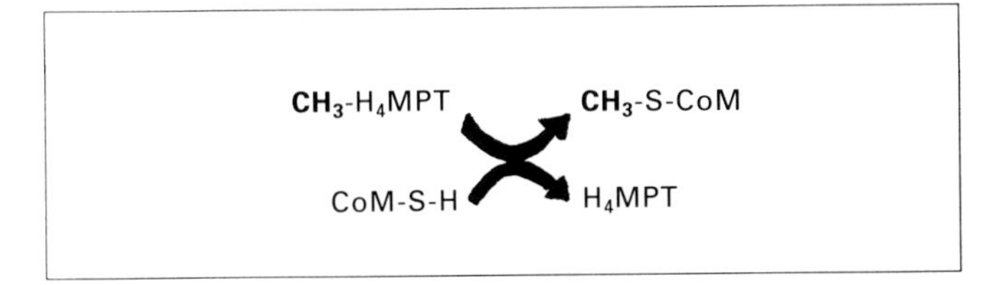

- Enzyme strongly membrane-bound
- Membranes of all methanogens contain high amounts of corrinoid protein (80 nmol/g of cells, 15 % of membrane protein)
- Corrinoid protein molecular mass ≈ 500 kDa
  8 corrinoids, 32 Fe/nS
  Subunits 33 kDa + Corrinoid
  15 kDa
  14 kDa
- Antibodies against 33 kDa protein cross-react with methyltransferase

Fig. 9. Some features of the methyltransferase system in methanogenic bacteria which catalyse the transfer of the methyl group from tetrahydromethanopterin to coenzyme M. This reaction is analogous to methionine synthesis from CH_3-tetrahydrofolate and homocysteine. The evidence that this enzyme is identical with the corrinoid membrane protein is still indirect.

tetrahydromethanopterin-bound methanol is almost at equilibrium; the enzymes mediating this part of the C_1 transformation sequence as well as the electron donors are probably all soluble.

(2) The reduction of methyltetrahydromethanopterin to methane requires two steps, the transfer of the C_1 unit to coenzyme M and finally the reduction of the coenzyme M-bound methanol to methane by mercaptoheptanoylthreoninephosphate; thereby a mixed heterodisulphide is formed. The former reaction is catalysed by a corrinoid-containing methyltransferase, the latter reaction is catalysed by the nickeltetrapyrrol-containing methyl-CoM reductase. All enzymes participating in this complex reaction may form a membrane-associated complex referred to as the methanoreductosome. The overall reaction may be associated with the translocation of possibly three to four Na^+; Na^+ is known to be required in the reduction of $-CH_3$ to CH_4. The sodium motive force is used to drive the initial reduction step. The exchange of the sodium and the proton motive force is thought to be catalysed by a sodium–proton antiporter.

The methyl transfer reaction between tetrahydromethanopterin and coenzyme M is still enigmatic (Fig. 9). Methanogens contain a membrane-bound corrinoid protein with a native molecular mass of 500 kDa [23–25]. The 500 kDa enzyme complex as isolated is very hydrophobic, it contains eight to nine corrinoids, four iron atoms per corrinoid, and an unidentified amount of acid-labile sulphur. The spectrum suggests that it contains iron–sulphur centres. This corrinoid protein is a dominant integral membrane protein in all methanogenic bacteria studied. It has recently been found that the enzyme that transfers the methyl group from tetrahydromethanopterin to coenzyme M is strictly membrane bound, that the methyltransferase contains a corrinoid, and that antibodies against methyltransferase react with the purified corrinoid membrane protein [25–29a]. Together, these facts are taken to indicate that the membrane-bound corrinoid protein complex is the actual methyltransferase and that the methyl transfer reaction may possibly be a Na^+ coupling site in methanogenesis. Yet, it is far from being clear how this reaction proceeds, why the protein is membrane bound, what the role of the iron–sulphur centre is, and how a transferase reaction should be energetically coupled.

(3) The reduction of the heterodisulphide to two thiols catalysed by heterodisulphide reductase is associated with the translocation of four H^+ (the Na^+ and H^+ stoichiometries refer to measurements with vesicles). Yet, there is no evidence so far that the heterodisulphide reductase is membrane bound. One may assume that the electron transfer from H_2 to the ultimate (soluble or membrane-associated) electron carrier, rather than the reduction of the heterodisulphide itself, drives the translocation of protons [9,10,14]. It is noteworthy that the actual net energy-yielding step in methanogenesis appears to be linked to sulphur rather than carbon reduction.

H^+-ATP synthase

It is a general belief that the proton motive force is coupled to the synthesis of ATP by an H^+-ATP synthase in all chemolithotrophic bacteria, except in those instances where a sodium motive force is the main driving force. This enzyme has not yet been characterized from sulphur-reducing or sulphate-reducing organisms, but only from methanogenic, sulphur-oxidizing, and halophilic archaebacteria. They contain a functional H^+-ATP synthase. This protein has a chimaeric structure in that the large F_1 subunits are similar to the V-type proteins, whereas the proteolipid of the F_0 part resembles the F-type ATP synthase. The role of a P-type ATPase found in a methanogen is not known.

Autotrophic CO_2 fixation in archaebacteria

CO_2 can be used as the sole carbon source for growth of most of the anaerobic archaebacteria isolated to date. The pathways of CO_2 fixation have recently been reviewed [1,6–8].

Reductive acetyl-CoA/CO dehydrogenase pathway

All autotrophic methanogens and the phylogenetically related autotrophic sulphate-reducing bacteria of the genus *Archaeoglobus* use a reductive acetyl-CoA pathway (Fig. 10). In brief, acetyl-CoA functions as the central biosynthetic intermediate; its formation from CO_2 requires the separate reduction of two

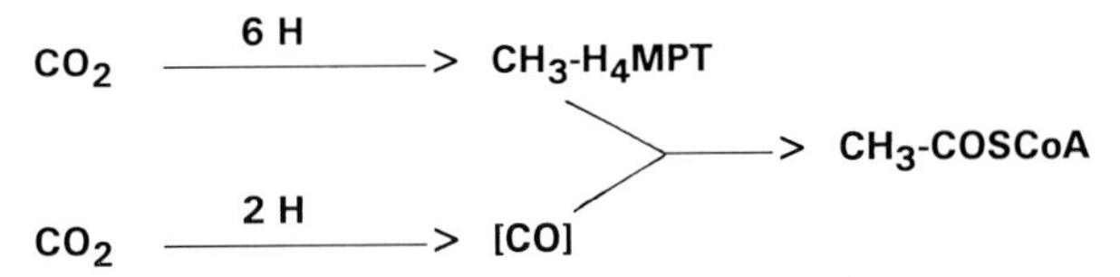

Fig. 10. Scheme of autotrophic CO_2 fixation by the reductive acetyl-CoA/CO-dehydrogenase pathway in methanogenic and sulphate-reducing archaebacteria. Abbreviation: H_4MPT, tetrahydromethanopterin. The reduction of CO_2 to CH_3–H_4MPT is shared with methanogenesis from CO_2.

molecules of CO_2, one to an enzyme-bound carbonyl group, the other to a tetrahydropterin-bound methyl-group. CO_2 reduction to the carbonyl, and condensation of the carbonyl unit and the methyl unit to acetyl-CoA, are both catalysed by the nickel enzyme complex CO dehydrogenase/acetyl-CoA synthase. The autotrophic pathway is very similar to the total synthesis of acetate from two CO_2 in *Clostridia* in which acetyl-CoA is an energy-rich intermediate. Acetyl-CoA is further reductively carboxylated to pyruvate. This reaction is common in most anaerobes which can use acetate as a main carbon source.

Reductive citric acid cycle

The sulphur-reducing archaebacterium *T. neutrophilus* uses a reductive citric acid cycle for CO_2 fixation [4,20,30–32] (Fig. 11). In brief, acetyl-CoA is made from two CO_2 by reverse reactions of the citric acid cycle which needs some modifications to be directed to the reductive side. The two carboxylating reactions are 2-oxoglutarate synthase (EC 1.2.7.3) and isocitrate dehydrogenase (EC 1.1.1.41). Notably, citrate synthase has to be replaced by ATP citrate lyase (EC 4.1.3.8) which generates oxaloacetate and releases acetyl-CoA. Unfortunately this enzyme has not yet been shown in this organism. Acetyl-CoA is reductively carboxylated to pyruvate by pyruvate synthase (EC 1.2.7.1).

Undefined reductive carboxylic acid cycles

In the aerobic archaebacterium *Sulfolobus brierleyi* evidence for a not yet defined reductive carboxylic acid cycle was obtained [33]. Whether this is a reductive citric acid cycle or not is unclear. There are facultatively anaerobic bacteria closely related to *Sulfolobus* which are not only able to oxidize sulphur to sulphate aerobically, but can also reduce sulphur to sulphide anaerobically. Under both growth conditions they grow autotrophically. It will be interesting to see whether they use one and the same autotrophic CO_2 fixation mechanism under both conditions. One wonders whether this pathway is related to the pathway in the sulphur-reducing, strictly anaerobic *Thermoproteus* which is only distantly related to the aerobic sulphur-oxidizing archaebacteria [34].

Calvin cycle

The key enzymes of the Calvin cycle, phosphoribulokinase (EC 2.7.1.19) and ribulose-1,5-bisphosphate carboxylase (EC 4.1.1.39), have been detected in several halophilic archaebacteria, albeit at very low specific activity [35]. This finding is

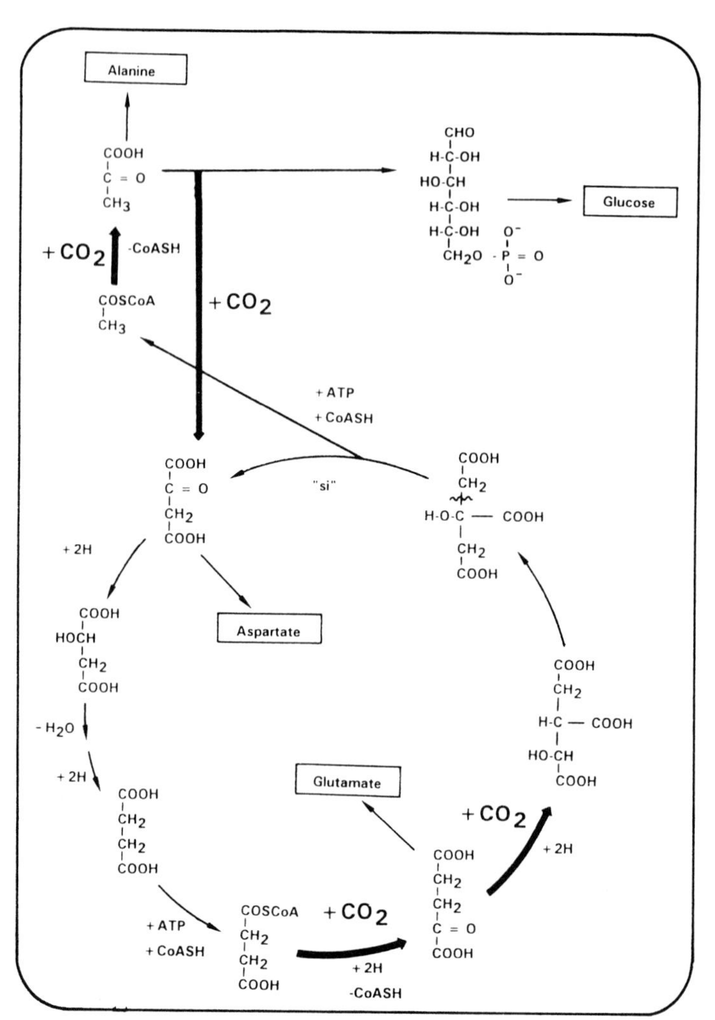

Fig. 11. Scheme of autotrophic CO_2 fixation by the reductive citric acid cycle in the sulphur-reducing archaebacterium *Thermoproteus neutrophilus*. Evidence for citrate cleavage to acetyl-CoA and oxaloacetate is indirect.

extremely interesting and needs to be studied in more detail. Especially interesting are the questions, (i) what is the function of the enzymes in these bacteria?, (ii) what is the degree of homology with known kinases and carboxylases from bacteria and plants? and (iii) were the proteins possibly acquired by lateral gene transfer from other bacteria? None of the halophilic archaebacteria has been shown to use CO_2 as the main carbon source. It has been proposed that the non-Calvin-type CO_2 fixation pathways evolved early in evolution and that the Calvin cycle was invented later.

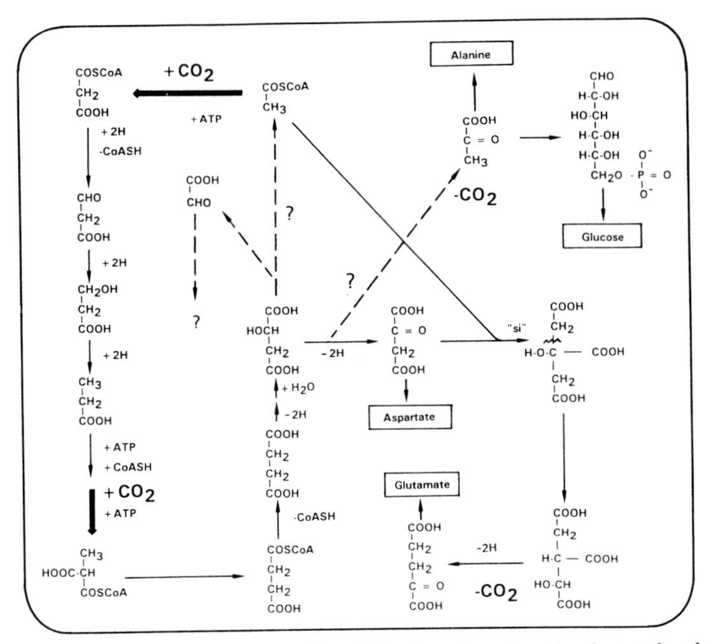

Fig. 12. Scheme of autotrophic CO_2 fixation by the hypothetical 3-hydroxypropionate cycle in *Chloroflexus aurantiacus*, a phototrophic eubacterium.

Autotrophic CO_2 fixation in the eubacterium *Chloroflexus aurantiacus*

So far there is no unique carbon fixation mechanism in archaebacteria which is not existent in eubacteria. However, CO_2 fixation has not been studied in all instances, and the CO_2 fixation pathway in the sulphur-oxidizing archaebacteria may be different from all known pathways. In this respect the emerging new CO_2 fixation cycle in the eubacterium *C. aurantiacus* may be of interest (Fig. 12). The evidence for a new pathway is convincing, although the data supporting a 3-hydroxypropionate cycle are circumstantial [32,36–39]. The organism is able to produce millimolar amounts of 3-hydroxypropionate from CO_2. 3-Hydroxypropionate as well as succinate are incorporated into all cellular compounds. Acetyl-CoA carboxylase and propionyl-CoA carboxylase activity were present. It is proposed that acetyl-CoA is carboxylated to malonyl-CoA, the activated carboxyl group is reduced to the alcohol function in 3-hydroxypropionate, and 3-hydroxypropionate is converted into propionyl-CoA. Propionyl-CoA carboxylation leads to methylmalonyl-CoA and from there malate and oxaloacetate are formed via succinate. It is postulated that malate is cleaved to regenerate acetyl-CoA and to release glyoxylate. The formation of cell material from glyoxylate is quite open. It may proceed via the glycerate or the serine pathway, as reported for other glyoxylate-utilizing organisms. A direct conversion of glyoxylate into acetyl-CoA seems unlikely. Pyruvate appears not to be formed by reductive carboxylation of acetyl-CoA.

References

1. Wood, H. G. (1990) in Autotrophic Bacteria (Schlegel, H. G. & Bowien, B., eds.), pp. 33–52, Science Tech Publishers, Madison
2. Blaut, M., Peinemann, S., Deppenmeier, U. & Gottschalk, G. (1990) FEMS Microbiol. Rev. **87**, 367–372
3. Daniels, L., Sparling, R. & Sprott, G. D. (1984) Biochim. Biophys. Acta **768**, 113–163
4. Danson, M. J. (1988) Adv. Microbiol. Physiol. **29**, 165–231
5. Di Marco, A. A., Bobik, T. A. & Wolfe, R. S. (1990) Annu. Rev. Biochem. **59**, 355–394
6. Diekert, G. (1990) FEMS Microbiol. Rev. **87**, 391–396
7. Fuchs, G. (1990) in Autotrophic Bacteria (Schlegel, H. G. & Bowien, B., eds.), pp. 365–384, Science Tech Publishers, Madison
8. Fuchs, G. (1990) in The Molecular Basis of Bacterial Metabolism (Hauska, G. & Thauer, R., eds.), pp. 13–20, Springer, Berlin
9. Gottschalk, G. (1990) in Autotrophic Bacteria (Schlegel, H. G. & Bowien, B., eds.), pp. 383–396, Science Tech Publishers, Madison
10. Gottschalk, G. & Blaut, M. (1990) Biochim. Biophys. Acta **1018**, 263–266
11. Jones, W. J., Nagle, D. P. & Whitman, W. B. (1987) Microbiol. Rev. **51**, 135–177
12. Keltjens, J. T., te Brömmelstroet, B. W., Kengen, S. W. M., van der Drift, C. & Vogels, G. D. (1990) FEMS Microbiol. Rev. **87**, 327–332
13. Kröger, A., Schröder, I., Krems, B. & Klimmek, O. (1990) in The Molecular Basis of Bacterial Metabolism (Hauska, G. & Thauer, R., eds.), pp. 128–133, Springer, Berlin
14. Müller, V., Blaut, M., Heise, R., Winner, C. & Gottschalk, G. (1990) FEMS Microbiol. Rev. **87**, 373–376
15. Thauer, R. K. (1990) Biochim. Biophys. Acta **1018**, 256–259
16. Thauer, R. K., Möller-Zinkhan, D. & Spormann, A. M. (1989) Annu. Rev. Microbiol. **43**, 43–67
17. Wolfe, R. S. (1985) Trends Biochem. Sci. **10**, 396–399
18. Wolfe, R. S. (1990) in The Molecular Basis of Bacterial Metabolism (Hauska, G. & Thauer, R., eds.), pp. 1–12, Springer, Berlin
19. Steudel, R. (1990) in Autotrophic Bacteria (Schlegel, H. G. & Bowien, B., eds.), pp. 289–304, Science Tech Publishers, Madison

19a. Krafft, T., Bokranz, M., Klimmek, O., Schröder, I., Fahrenholz, F., Kojro, E. & Kröger, A. (1992) Eur. J. Biochem. in the press

19b. Dahl, C., Koch, H. G., Keuken, O. & Trüper, H. G. (1990) FEMS Microbiol. Lett. **67**, 27–32

19c. Lampreia, J., Fauque, G., Speich, N., Dahl, C., Moura, I., Trüper, H. G. & Moura, J. J. G. (1991) Biochem. Biophys. Res. Commun. **181**, 342–347

20. Schäfer, S., Barkowski, C. & Fuchs, G. (1986) Arch. Microbiol. **146**, 301–308
21. Thauer, R. K. (1990) in Autotrophic Bacteria (Schlegel, H. G. & Bowien, B., eds.), pp. 397–413, Science Tech Publishers, Madison
22. Kaesler, B. & Schönheit, P. (1989) Eur. J. Biochem. **186**, 309–316
23. Dangel, W., Schulz, H., König, H. & Fuchs, G. (1987) Arch. Microbiol. **148**, 52–56
24. Schulz, H., Albracht, S. P. J., Coremans, J. M. C. C. & Fuchs, G. (1988) Eur. J. Biochem. **177**, 589–597
25. Stupperich, E., Juza, A., Eckerskorn, C. & Edelmann, L. (1990) Arch. Microbiol. **155**, 28–34
26. Kengen, S. W. M., Daas, P. J. H., Keltjens, J. T., van der Drift, C., Vogels, G. D. (1990) Arch. Microbiol. **154**, 156–161
27. Kengen, S. W. M., Mosterd, J. J., Nelissen, L. H., Keltjens, J. T., van der Drift, C. & Vogels, G. D. (1988) Arch. Microbiol. **150**, 405–412

28. Van de Wijngaard, W. M. H., Creemers, J., Vogels, G. D. & Van der Drift, C. (1991) FEMS Microbiol. Lett. **80**, 207–212
29. Van de Wijngaard, W. M. H., van der Drift, C. & Vogels, G. D. (1988) FEMS Microbiol. Lett. **52**, 165–172
29a. Fischer, R., Gärtner, P. & Thauer, R. K. (1992) Arch. Microbiol. in the press
30. Schäfer, S., Götz, M., Eisenreich, W., Bacher, A. & Fuchs, G. (1989) Eur. J. Biochem. **184**, 151–156
31. Schäfer, S., Paalme, T., Vilu, R. & Fuchs, G. (1989) Eur. J. Biochem. **186**, 695–700
32. Strauß, G., Eisenreich, W., Bacher, A. & Fuchs, G. (1992) Eur. J. Biochem. in the press
33. Kandler, O. & Stetter, K. O. (1981) Zentralbl. Bakteriol. Abt. I. Orig. C**2**, 111–121
34. Brock, T. D. (1990) in Autotrophic Bacteria (Schlegel, H. G. & Bowien, B., eds.), pp. 499–512, Science Tech Publishers, Madison
35. Rawal, N., Kelkar, S. M. & Altekar, W. (1988) Biochem. Biophys. Res. Commun. **156**, 451–456
36. Holo, H. & Grace, D. (1987) Arch. Microbiol. **148**, 292–297
37. Holo, H. & Sirevåg, R. (1986) Arch. Microbiol. **145**, 173–180
38. Holo, H. (1986) PhD thesis, University of Oslo, Oslo
39. Holo, H. (1989) Arch. Microbiol. **151**, 252–256

Biochem. Soc. Symp. **58**, 41–49
Printed in Great Britain

Biochemistry of methanogenesis

Ralph S. Wolfe

Department of Microbiology, University of Illinois, Urbana, IL 61801, U.S.A.

Introduction

One of nature's most interesting enzymic reactions is the reduction of a methyl group to methane. The methylreductase reaction on a global scale is estimated to produce 400 metric tons of CH_4/year. The reaction responsible for this vast output is found in a group of methanogenic archaebacteria (archaea) known as the methanogens. These organisms constitute a major division of the archaea and represent an ancient divergence in evolution. The methylreductase reaction is confined to the methanogens and has not been found elsewhere in nature. Reduction of a methyl group to methane is an energetically favourable, essentially irreversible reaction, $\Delta G^{0\prime} = -45$ kJ/mol of CH_4 [1]. Methanogens are highly specialized, and all methanogens so far examined employ a methyl group as their terminal electron acceptor. This reaction is obligatory for growth; no investigator has succeeded in growing these organisms in any other way. Methanogens oxidize a narrow range of substrates, the major ones being hydrogen, formate, methanol, methylamines and acetate. Despite the specialized niche which the methanogens occupy their habitats are diverse and range from submarine hydrothermal vents, (where extreme thermophiles such as *Methanopyrus kandleri*, 110 °C, or *Methanococcus jannaschii*, 85 °C, grow on the hydrogen and carbon dioxide in volcanic gases) to aquatic sediments, bogs and swamps, and to the digestive tracts of animals, especially the delayed-passage organs of cellulose digesters (rumen, caecum, and hind gut of certain termites).

We chose the thermophile, *Methanobacterium thermoautotrophicum*, for our study of the biochemistry of methanogenesis. The organism grows at 65 °C on hydrogen and reduces carbon dioxide to methane, readily producing kilogram quantities of cells; the enzymes are stable to fractionation at room temperature. The review of DiMarco *et al.* may be consulted for a more detailed account of the coenzymes of methanogens [2].

$HS-CH_2-CH_2-SO_3^-$

Coenzyme M (HS-CoM)

$CH_3-S-CH_2-CH_2-SO_3^-$

$CH_3-S-CoM$

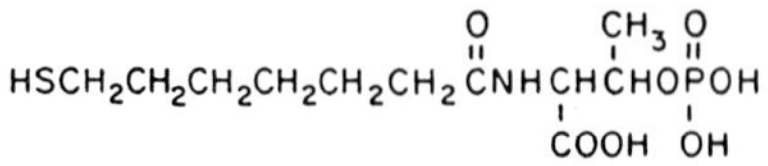

Factor F_{430}

$HSCH_2CH_2CH_2CH_2CH_2CH_2C(O)NHCH(COOH)CH(CH_3)OP(O)(OH)OH$

7-Mercaptoheptanoylthreonine phosphate (HS-HTP)

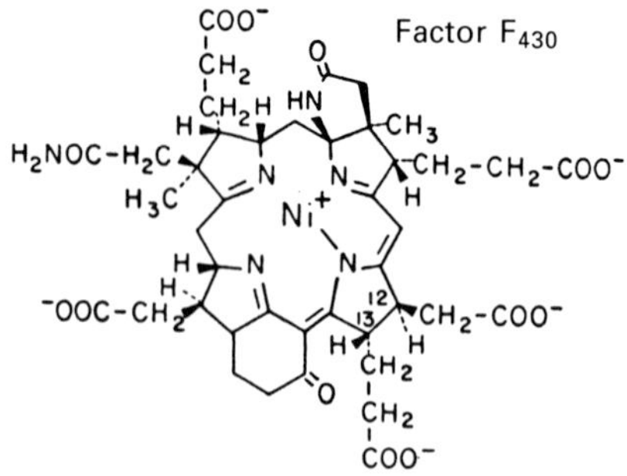

$$CH_3\text{-S-CoM} + \text{HS-HTP} \xrightarrow{\text{Methylreductase}} CH_4 + \text{CoM-S-S-HTP}$$

Enzyme: $M_r = 300\,000\ \alpha_2\beta_2\gamma_2$

$\alpha = 66\,000 \quad \beta = 48\,000 \quad \gamma = 37\,000$

2 mol F_{430}/mol of protein

Fig. 1. The methylreductase and its unique coenzymes.

The methylreductase

Although the reduction of a methyl group to methane appeared to be a straightforward reaction, elucidation of the methylreductase reaction required a 25-year effort. The first methyl group to be identified as a substrate was the methyl group of methylcobalamin, which was readily converted into methane by cell extracts [3]. With the discovery of 2-mercaptoethanesulphonic acid (coenzyme M, HS-CoM) and its requirement for the methylreductase reaction [4] focus shifted to 2-(methyl-thio)ethanesulphonic acid (CH_3-S-CoM) as the true substrate (Fig. 1) for the methylreductase [5]. A methyltransferase which transferred the methyl group from methylcobalamin to HS-CoM was highly purified [6]. The CH_3-S-CoM reductase was fractionated into three components [7], each of which, in addition to ATP, was required to reconstitute the methylreductase reaction. Component A consisted of a fraction of high-molecular-mass proteins which contained hydrogenase activity. Component B was a small dialysable coenzyme. Component C was an oxygen-stable protein, which when purified to homogeneity was yellow. However, these preparations retained less than 1 % of the activity of whole cells. A non-fluorescent, yellow compound had been found previously in cell extracts [8] and named factor 430 (F_{430}) because of its strong absorption at 430 nm; its function was unknown. F_{430} was found to contain nickel [9,10], and in the laboratory of R. K. Thauer evidence suggested that F_{430} could be a tetrapyrrole [11]. The structure of F_{430} (Fig. 1), the first nickel-tetrapyrrole, a tetrahydrocorphin, was determined in Eschenmoser's laboratory [12]. The yellow chromophore of component C was found to be F_{430}, and the incorporation of ^{63}Ni into the enzyme was documented [13]. It was suspected that component C was the methylreductase, but this was difficult to prove, since the A proteins were required for the methylreductase reaction [14]. However, Ankel-Fuchs in Thauer's laboratory showed that component C was the methylreductase. She was able to prepare component C which did not require the A proteins or ATP for methylreductase activity, suggesting that the A proteins and ATP were involved in converting inactive component C into the active form [15,16] where Ni(I) was believed to be the active state of nickel.

But, what was component B? Fractionation and purifications were carried out in several laboratories, and in my own laboratory a ten-year effort resulted in a structure and synthesis of a compound that functioned as component B, 7-mercaptoheptanoylthreonine phosphate (HS-HTP), Fig. 1 [17,18]. When this compound was tested with an active preparation of component C, ATP was not required for methane formation [19]. It was now certain that CH_3-S-CoM was the substrate, component C was the methylreductase, and HS-HTP was the electron donor. But what was the product? As shown by the reaction in Fig. 1 a heterodisulphide (CoM-S-S-HTP) formed between the methyl-donating and the electron-donating coenzyme was the product [20,21]. Recently, very active preparations of CH_3-S-CoM reductase have been prepared in Thauer's laboratory [22] with HS-HTP as the electron donor. Evidence has been presented for localization of the methylreductase in a structure named the 'methanoreductosome' [23], attached to the cell membrane. Recently, a membrane fraction which contains the methylreductase has been shown to be more active with a factor named the methyl-reducing factor [24]. The structure of this factor, which has been shown to contain HS-HTP, is uridine 5′-[*N*-(7-mercaptoheptanoyl)-*O*-3-phosphothreonine-P-yl(2-acetamido-2-deoxy-β-mannopyranuronsyl) (acid anhydride)]-(1 → 4)-*O*-2-acetamido-2-deoxy-α-glucopyranosyl diphosphate. HS-HTP is linked to C-6 of 2-acetamido-2-deoxymannopyranuronic acid of the UDP-disaccharide through a carboxylic–phosphoric anhydride linkage. This bond may be responsible for the instability of the molecule during isolation. Possibly the UDP-disaccharide serves to anchor the coenzyme in the methanoreductosome. The proposed abbreviation MRF for this coenzyme is unfortunate, for it causes confusion with MFR, an established abbreviation for methanofuran. It is suggested here that coenzyme B be used as an appropriate abbreviation for the complete structure, HS-CoB being the abbreviation for the active form. HS-HTP would be used, when only this portion of the coenzyme is employed in reaction mixtures.

Coenzymes of C_1 transfer and reduction from CO_2 to the methyl level

Although the archaebacterial coenzymes of C_1 transfer and reduction were discovered in the chronological order of coenzyme F_{420}, tetrahydromethanopterin and methanofuran, their sequential roles in CO_2 reduction are presented in Fig. 2. Before 1978 the only evidence concerning CO_2 activation and reduction to CH_4 in methanogens was the stimulation of methane formation from CO_2 in cell extracts by the addition of CH_3-S-CoM [25]. This phenomenon was named the RPG effect after R. P. Gunsalus who discovered it. We concluded that the CH_3-S-CoM reductase was in some way connected to CO_2 activation. In his thesis in 1978, J. A. Romesser reported the first evidence for a new cofactor that greatly stimulated CO_2 conversion

Methanofuran (MFR)

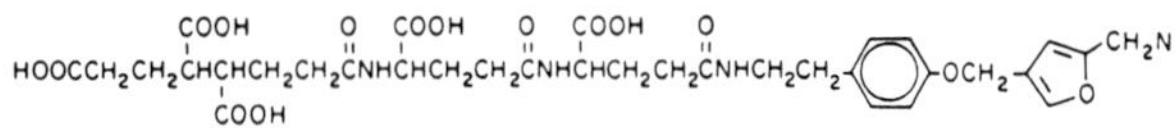

1. $CO_2 + MFR + 2H \xrightarrow{CO_2\ reductase} Formyl\text{-}MFR$

2. $Formyl\text{-}MFR + H_4MPT \xrightarrow{Formyltransferase} Formyl\text{-}H_4MPT + MFR$

Tetrahydrometrhanopterin (H_4MPT)

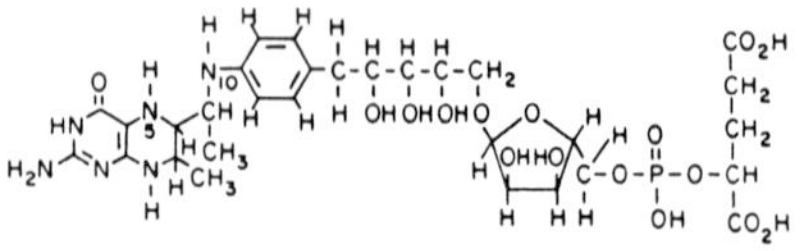

3. $Formyl\text{-}H_4MPT \xrightarrow{Cyclohydrolase} Methenyl\text{-}H_4MPT + H_2O$

Coenzyme F_{420}

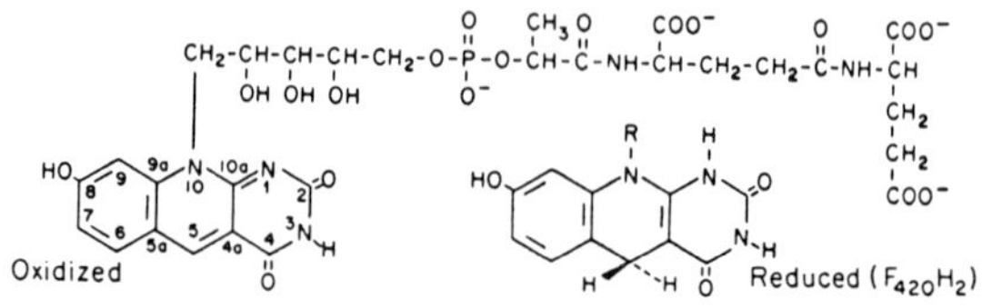

Oxidized

Reduced ($F_{420}H_2$)

4. $H_2 + F_{420} \xrightarrow{Hydrogenase} F_{420}H_2$

5a. $Methenyl\text{-}H_4MPT + F_{420}H_2 \xrightarrow{Oxidoreductase} Methylene\text{-}H_4MPT + F_{420}$

5b. $Methenyl\text{-}H_4MPT + H_2 \xrightarrow{Hydrogenase} Methylene\text{-}H_4MPT$

6. $Methylene\text{-}H_4MPT + F_{420}H_2 \xrightarrow{Reductase} Methyl\text{-}H_4MPT + F_{420}$

Fig. 2. Coenzymes of C_1 transfer and the enzyme reactions in which they participate. Abbreviations: MFR, methanofuran; $F_{420}H_2$, reduced factor 420.

into methane. We named this compound the carbon dioxide reduction (CDR) factor [26]. Cell extract was resolved for this factor by passage through a short G-25 Sephadex column. When John Leigh continued the study of the CDR factor, he found that active fractions consisted of two factors, one of which was the CDR factor [27], the other being methanopterin. The CDR factor was purified to homogeneity and its structure was determined [28]. This long linear molecule with hydrophilic groups at one end and a furan group with a primary amine at the other was named methanofuran, Fig. 2. The coenzyme was found to be formylated on the primary amine of the furan ring, formylmethanofuran being the first stable product of CO_2 fixation [29]. The electron donor for this reaction (Fig. 2, reaction 1) has not yet been determined. The reaction has been studied in the reverse direction, e.g. formylmethanofuran dehydrogenase by the use of dyes as electron acceptors [30]. At this time it has not been unequivocally established that the CO_2 reductase and the dehydrogenase are the same protein.

The formyl group of formylmethanofuran is transferred by a formyltransferase to a pterin (Fig. 2, reaction 2). The pterin was first noticed as a bright blue fluorescent band on chromatographic columns and was named F_{342} for its absorbance maximum [8]. The compound was established as a 7-methylpterin [31]. Escalante-Semerena discovered a factor in cell extracts that was required for the conversion of formaldehyde into methane, and he named this compound FAF, the formaldehyde activation factor [32]. He showed that it had properties similar to methanopterin, MPT, [33] and documented that the molecular mass of the reduced form was 776 Da and the oxidized form was 772 Da; FAF was tetrahydromethanopterin (H_4MPT). The structure proved difficult to determine, and was only established by the definitive two dimensional n.m.r. work of van Beelen *et al.* on another compound, the yellow fluorescent compound (YFC) of Daniels. This compound was determined to be 5,10-methenyltetrahydromethanopterin [34]. So the terms, F_{342}, FAF and YFC found in the literature are no longer used; MPT is the oxidized form and H_4MPT (5,6,7,8-tetrahydromethanopterin) is the active form of the coenzyme which functions in the archaebacteria as a C_1 carrier, in a role similar to that of folate in prokaryotes and eukaryotes [35]. However, there are unique differences; H_4MPT is formylated by a formyltransferase, an enzyme foreign to folate biochemistry. The product of the formyltransferase (Fig. 2, reaction 2) is 5-methyl-H_4MPT [36]. This enzyme has been cloned, sequenced, and expressed in *Escherichia coli* [37], the expressed enzyme being active at 60 °C. In a reaction similar to that of folates (Fig. 2, reaction 3) water is removed from the formyl group by a cyclohydrolase to form 5,10-methenyl-H_4MPT [38]. The methenyl derivative absorbs strongly at 335 nm and enzymic reactions involving this compound may easily be followed at this wavelength [35].

In the late 1960s Paul Cheeseman discovered a yellow compound in cell extracts that exhibited a bright blue-green fluorescence. He named this compound factor 420 (F_{420}) for its strong absorption in the oxidized state at 420 nm [39]. This absorption is pH dependent, the isosbestic point being 401 nm [40] with an absorption coefficient of 25.9 $mM^{-1} \cdot cm^{-1}$. Dudley Eirich determined the structure of F_{420} (Fig. 2) as an 8-hydroxy-5-deazaflavin derivative [41]. F_{420} was found to be a coenzyme for formate dehydrogenase [42] and for hydrogenase [43]. A specific F_{420}-reducing hydrogenase was purified from *Methanobacterium* [44], that effects reaction 4 (Fig. 2). F_{420} is a major electron carrier in methanogens as well as in other archaebacteria. Methenyl-H_4MPT is reduced to methylene-H_4MPT by an F_{420} oxidoreductase, Fig. 2, reaction 5*a* [45] or by a unique small hydrogenase [30], reaction 5*b*.

The methylene-H_4MPT reductase (Fig. 2, reaction 6) also has been shown to employ reduced F_{420} as the electron donor [46], the product being 5-methyl-H_4MPT. So the major C_1 carrier in methanogenesis is H_4MPT which carries a C_1 group at the formyl, methenyl, methylene, and methyl states.

It is proposed that there are two methyltransferase steps between methyl-H_4MPT and HS-CoM:

$$\text{Methyl-}H_4\text{MPT} + \text{Cobamide-enzyme} \xrightarrow{\text{Methyltransferase 1}} \text{Methyl-cobamide enzyme} + H_4\text{MPT} \quad (7)$$

$$\text{Methyl-cobamide-enzyme} + \text{HS-CoM} \xrightarrow{\text{Methyltransferase 2}} CH_3\text{-S-CoM} + \text{Cobamide enzyme} \quad (8)$$

Although reaction (7) has not been extensively studied in different methanogens transfer of the methyl group from methyl-H_4MPT to the cobalt atom of a cobamide enzyme is believed to be the first step in methyl transfer. In the second step (reaction 8), transfer of the methyl group from the cobalt atom to HS-CoM is believed to involve the methyltransferase described some years ago [6].

The link between the methylreductase (Fig. 1) and CO_2 activation (reaction 1, Fig. 2) was found to be CoM-S-S-HTP; the addition of the heterodisulphide to reaction mixtures stimulated formylmethanofuran synthesis from CO_2 40-fold [47]. Although the mechanism of this reaction remains obscure, it has been proposed that CoM-S-S-HTP activates an unknown, low-potential electron carrier that is involved in CO_2 reduction [48]. Regeneration of the free thiols from the heterodisulphide may occur by an $F_{420}H_2$ (reduced factor 420)-driven heterodisulphide reductase [50] or by an F_{420}-independent membrane-catalysed reaction [51,52].

Nature's strategies

Nature has evolved three major strategies for methanogenesis (Fig. 3), each employing the central role of the terminal methylreductase and the heterodisulphide reductase reactions. In strategy I (Fig. 3) CO_2 is the electron sink, the molecule being reduced to the methyl level by reactions shown in Fig. 2. Hydrogen is the major electron-donor substrate. This is the most commonly used strategy, and only a few species are unable to oxidize hydrogen and reduce CO_2. Formate is used as a substrate by many methanogens where it is first oxidized to hydrogen and carbon dioxide. A few species have been shown to oxidize primary and secondary alcohols. For example, 2-propanol is oxidized to acetone, and the electrons enter the CO_2 reduction pathway [53].

In strategy II pre-formed methyl groups from methanol or methylamines are transferred to HS-CoM; CH_3-S-CoM becomes the electron acceptor. Another methyl group from methanol or methylamine is activated and oxidized to CO_2 via the reversal of the pathway shown in Fig. 2, formylmethanofuran dehydrogenase being the terminal reaction. Thus, the methyl groups from three CH_3OH molecules serve as electron acceptors for the six electrons generated by the oxidation of one CH_3OH to CO_2. In variations on this theme at least one organism has been described that can oxidize hydrogen and reduce the methyl group of methanol to methane [54].

In strategy III (Fig. 3) acetate is the sole substrate, and only a few methanogens are able to make a living on this energetically poor reaction ($\Delta G^{0\prime} = -36$ kJ/mol). Methanogenesis from acetate is a major source of methane in sludge digesters. Acetate is activated to acetylphosphate by an ATP-driven acetate kinase, and acetyl-CoA is then formed by a phosphotransacetylase; acetyl-CoA is the substrate for carbon monoxide dehydrogenase (CODH), the key enzyme in methanogenesis from acetate.

The CODH enzyme complex is composed of five subunits [55] and has been resolved into two components; one is a nickel-containing iron–sulphur protein that contains CODH activity, and is believed to be the site of carbon–carbon bond cleavage as well as carbon–sulphur bond cleavage of acetyl-CoA. Thus, a methyl group, a carbonyl group, and HS-CoA would be formed by the nickel/iron–sulphur protein. A second component is a corrinoid-containing iron–sulphur protein. It is believed that this protein accepts the methyl group generated by the nickel iron–sulphur protein and donates it either to form 5-methyl-H_4MPT or directly to HS-CoM. A ferredoxin has been shown to be the electron acceptor from oxidation of the carbonyl group to CO_2 [55]. Methanogens that carry out methanogenesis from

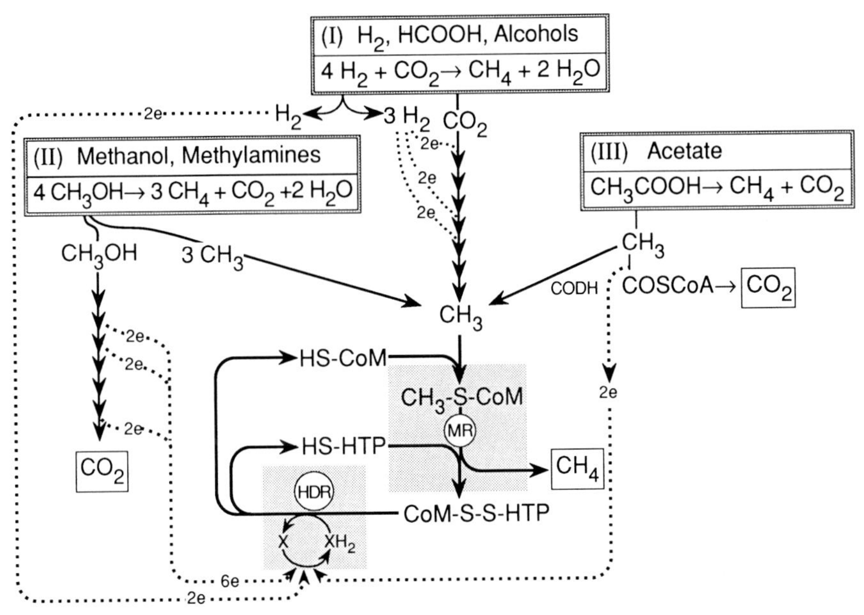

Fig. 3. The central role of the methylreductase and heterodisulphide reductase in methanogenesis from the three major substrate sources. (I) Reduction of CO_2, (II) reduction of a preformed methyl group, and (III) oxidation of acetate. Abbreviations used: MR, methylreductase; HDR, heterodisulphide reductase.

acetate possess cytochrome *b*. The route of electrons from CODH to the heterodisulphide reductase is not understood at present. Perhaps it is direct from ferredoxin or it may also involve cytochrome *b* in the membrane.

ATP

For many years it seemed possible that substrate-level phosphorylation could occur in methanogens, since perhaps energy could be conserved from the methylreductase reaction. However, no substrate-level phosphorylation has been found, and evidence is accumulating that anaerobic respiration is the major theme of the methanogens. With the discovery of CoM-S-S-HTP as the product of the methylreductase reaction, a keen interest was awakened in proton-driven bioenergetics. Reduction of CoM-S-S-HTP consumes two protons and two electrons in generating HS-CoM and HS-HTP with a stoichiometry of one ATP synthesized per two CoM-S-S-HTP reduced [50]. For example, inside-out vesicles in the presence of a membrane-located hydrogenase and a heterodisulphide reductase, located on the outside of the membrane, accumulated protons inside the vesicles as electrons and protons were used by the external heterodisulphide reductase [49]. Recently, an $F_{420}H_2$-driven heterodisulphide reductase has been documented [50]. Although the answers are not complete a sodium pump has been implicated to operate between methylene-H_4MPT and CH_3-S-CoM. It appears that energy conservation in methanogens involves both proton and sodium pumps. The energy metabolism of methanogenic bacteria has been recently reviewed by Thauer [1].

References

1. Thauer, R. K. (1990) Biochim. Biophys. Acta **1018**, 256–259
2. DiMarco, A. A., Bobik, T. A. & Wolfe, R. S. (1990) Annu. Rev. Biochem. **59**, 355–394
3. Blaylock, A. B. & Stadtman, T. C. (1963) Biochem. Biophys. Res. Commun. **11**, 34–38
4. McBride, B. C. & Wolfe, R. S. (1971) Biochemistry **10**, 2317–2324
5. Taylor, C. D. & Wolfe, R. S. (1974) J. Biol. Chem. **249**, 4879–4885
6. Taylor, C. D. & Wolfe, R. S. (1974) J. Biol. Chem. **249**, 4886–4890
7. Gunsalus, R. P. & Wolfe, R. S. (1980) J. Biol. Chem. **255**, 1891–1895
8. Gunsalus, R. P. & Wolfe, R. S. (1978) FEMS Microbiol. Lett. **3**, 191–193
9. Diekert, G., Jaenchen, R. & Thauer, R. K. (1980) FEBS Lett. **119**, 118–120
10. Whitman, W. B. & Wolfe, R. S. (1980) Biochem. Biophys. Res. Commun. **92**, 1196–1201
11. Diekert, G., Klee, B. & Thauer, R. K. (1980) Arch. Microbiol. **124**, 103–106
12. Livingston, D. A., Pfaltz, A., Schreiber, J., Eschenmoser, A. & Ankel-Fuchs, D. (1984) Helv. Chim. Acta **67**, 334–351
13. Ellefson, W. L. & Wolfe, R. S. (1981) J. Biol. Chem. **256**, 4259–4262
14. Nagle, D. P., Jr. & Wolfe, R. S. (1983) Proc. Natl. Acad. Sci. U.S.A. **80**, 2151–2155
15. Ankel-Fuchs, D., Hüster, R., Mörschel, E., Albracht, S. P. J. & Thauer, R. K. (1986) Syst. Appl. Microbiol. **7**, 383–387
16. Rouviére, P. E. & Wolfe, R. S. (1988) J. Biol. Chem. **263**, 7913–7916
17. Noll, K. M., Rinehart, K. L., Jr., Tanner, R. S. & Wolfe, R. S. (1986) Proc. Natl. Acad. Sci. U.S.A. **83**, 4238–4242
18. Noll, K. M., Donnelly, M. I. & Wolfe, R. S. (1987) J. Biol. Chem. **262**, 513–515
19. Ankel-Fuchs, D., Böcher, R., Thauer, R. K., Noll, K. M. & Wolfe, R. S. (1986) FEBS Lett. **213**, 123–127
20. Bobik, T. A., Olson, K. D., Noll, K. M. & Wolfe, R. S. (1987) Biochem. Biophys. Res. Commun. **149**, 455–460
21. Ellermann, J., Hedderich, R., Böcher, R. & Thauer, R. K. (1988) Eur. J. Biochem. **171**, 669–677
22. Ellermann, J., Rospert, S., Thauer, R. K., Bokranz, M., Klein, A. & Voges, M. (1989) Eur. J. Biochem. **184**, 63–68
23. Mayer, F., Rohde, M., Salzmann, M., Jussofie, A. & Gottschalk, G. (1988) J. Bacteriol. **170**, 1438–1444
24. Sauer, F. D., Blackwell, B. A., Kramer, J. K. G. & Marsden, B. J. (1990) Biochemistry **29**, 7593–7600
25. Gunsalus, R. P. & Wolfe, R. S. (1978) Biochem. Biophys. Res. Commun. **76**, 790–795
26. Romesser, J. A. (1978) Ph.D. thesis, Urbana, University of Illinois
27. Leigh, J. A. & Wolfe, R. S. (1983) J. Biol. Chem. **258**, 7435–7440
28. Leigh, J. A., Rinehart, K. L., Jr. & Wolfe, R. S. (1984) J. Am. Chem. Soc. **106**, 3636–3640
29. Leigh, J. A., Rinehart, K. L., Jr. & Wolfe, R. S. (1985) Biochemistry **24**, 995–999
30. Schworer, B. & Thauer, R. K. (1991) Arch. Microbiol. **155**, 459–465
31. Keltjens, J. T., Huberts, M. J., Laarhoven, W. H. & Vogels, G. D. (1983) Eur. J. Biochem. **130**, 537–544

32. Escalante-Semerena, J. C., Leigh, J. A., Rinehart, K. L., Jr & Wolfe, R. S. (1984) Proc. Natl. Acad. Sci. U.S.A. **81**, 1976–1980
33. Vogels, G. D., Keltjens, J. T., Hutten, T. J. & Van der Drift, C. (1982) Zentralbl. Bakteriol. Mikrobiol. Hyg. Abt. I Orig. C **3**, 258–264
34. van Beelen, P., Stassen, A. P. M., Bosch, J. W. G., Vogels, G. D., Guijt, W. & Hassnoot, C. A. G. (1984) Eur. J. Biochem. **138**, 563–571
35. Escalante-Semerena, J. C., Rinehart, K. L., Jr. & Wolfe, R. S. (1984) J. Biol. Chem. **259**, 9447–9455
36. Donnelly, M. I., Escalante-Semerena, J. C., Rinehart, K. L., Jr. & Wolfe, R. S. (1985) Arch. Biochem. Biophys. **242**, 430–439
37. DiMarco, A. A., Sment, K. A., Konisky, J. & Wolfe, R. S. (1990) J. Biol. Chem. **265**, 472–476
38. DiMarco, A. A., Donnelly, M. I. & Wolfe, R. S. (1986) J. Bacteriol. **168**, 1372–1377
39. Cheeseman, P., Toms-Wood, A. & Wolfe, R. S. (1972) J. Bacteriol. **112**, 527–531
40. Eirich, L. D., Vogels, G. D. & Wolfe, R. S. (1979) J. Bacteriol. **140**, 20–27
41. Eirich, L. D., Vogels, G. D. & Wolfe, R. S. (1978) Biochemistry **17**, 4583–4593
42. Tzeng, S. F., Bryant, M. P. & Wolfe, R. S. (1975) J. Bacteriol. **121**, 192–196
43. Tzeng, S. F., Wolfe, R. S. & Bryant, M. P. (1975) J. Bacteriol. **121**, 184–190
44. Jacobson, S. F., Daniels, L., Fox, J. A., Walsh, C. T. & Orme-Johnson, W. H. (1982) J. Biol. Chem. **257**, 3385–3388
45. te Brommelstroet, B. W. T., Hensgens, C. M. H., Keltjens, J. T., Vanderdrift, C. & Vogels, G. D. (1991) Biochim. Biophys. Acta **1073**, 77–84
46. Ma, K. & Thauer, R. K. (1990) Eur. J. Biochem. **191**, 187–193
47. Bobik, T. A. & Wolfe, R. S. (1988) Proc. Natl. Acad. Sci. U.S.A. **85**, 60–63
48. Bobik, T. A. & Wolfe, R. S. (1989) J. Bacteriol. **171**, 1423–1427
49. Pinemann, S., Hedderich, R., Blaut, M., Thauer, R. K. & Gottschalk, G. (1990) FEBS Lett. **263**, 57–60
50. Deppenmeier, U., Blaut, M., Mahlmann, A. & Gottschalk, G. (1990) Proc. Natl. Acad. Sci. U.S.A. **87**, 9449–9453
51. Hedderick, R. & Thauer, R. K. (1988) FEBS Lett. **234**, 223–227
52. Deppenmeier, U., Blaut, M. & Gottschalk, G. (1991) Arch. Microbiol. **155**, 272–277
53. Widdel, F. & Wolfe, R. S. (1989) Arch. Microbiol. **152**, 322–328
54. Miller, T. L. & Wolin, M. J. (1985) Arch. Microbiol. **151**, 116–122
55. Albanat, D. R. & Ferry, J. G. (1991) Proc. Natl. Acad. Sci. U.S.A. **88**, 3272–3276

Biochem. Soc. Symp. **58**, 51–72
Printed in Great Britain

Archaebacterial lipids: structure, biosynthesis and function

Morris Kates

Department of Biochemistry, University of Ottawa, Ottawa, ON, Canada, K1N 6N5

Introduction

The archaebacteria [1] (now designated Archaea [2]) are recognized as a separate domain of ancient organisms that have evolved under the conditions of the primitive earth, such as high temperature, anaerobic atmosphere and high salinity. Three groups of archaebacteria exist: the extreme halophiles, the methanogens and the thermoacidophiles. They are clearly distinguished from all other organisms, particularly the eubacteria, by three major criteria: (1) their 16S ribosomal RNA sequences differ distinctly from those of eubacteria and eukaryotes; (2) their cell walls consist of glycosylated proteins rather than the peptidoglycan structure in eubacteria; and (3) their membrane lipids are unique in consisting entirely of derivatives of a C_{20}–C_{20} diacylglycerol diether: *sn*-2,3-diphytanylglycerol diether (structure **1**, Fig. 1) [3], and its dimer, dibiphytanyldiglycerol tetraether (structure **2**, Fig. 2) [4], in which two C_{20}–C_{20} diether moieties are linked head-to-head. The configuration of the two glycerols in the tetraether is the same as in the diether (**1**) [5,6]. Also, the phytanyl groups have the configuration 3*R*, 7*R*, 11*R* [3] and the biphytanyl groups in the glycerol tetraether have the configuration 3*R*, 7*R*, 11*R*, 15*R*, 15′*R*, 11′*R*, 7′*R*, 3′*R* [7].

Variants of the diphytanylglycerol diether (**1**) and the dibiphytanyldiglycerol tetraether (**2**) core lipid structures are also found in some archaea. For example, both the C_{20}–C_{25} diether (structure **1a**, Fig. 1) and C_{25}–C_{25} diether (structure **1b**) core lipids occur in alkaliphilic species of extreme halophiles (*Natronobacterium* and *Natronococcus*) [8,9], and the C_{20}–C_{25} diether also occurs in Halococci [10]; a macrocyclic C_{40}-diether (structure **1c**, Fig. 1) [11] is the major core lipid in the thermophilic methanogen *Methanococcus jannaschii*; and *Methanosarcina* and *Methanosaeta* (*Methanothrix*) species contain novel 3-hydroxydiether core lipids (structures **1d** and **1e**, respectively) [12,13]. Species of the thermoacidophilic *Sulfolobus* genus contain lipids derived from a dibiphytanylglycerol nonitoltetraether (structure

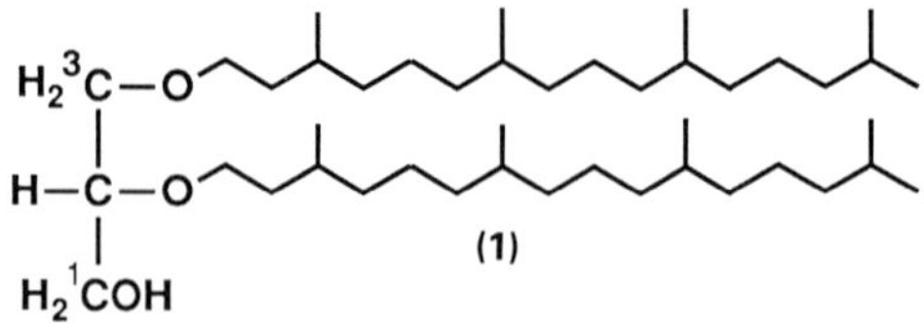

Fig. 1. Diphytanylglycerol ether (1, archaeol) and derivatives (1a–1e, see text). These form the core lipids of extreme halophiles and some methanogens.

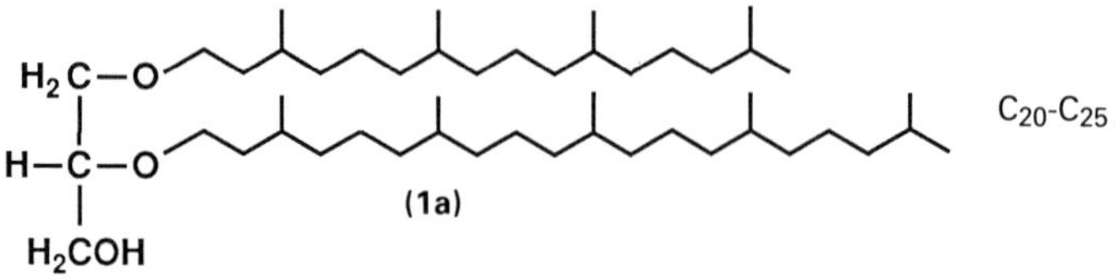

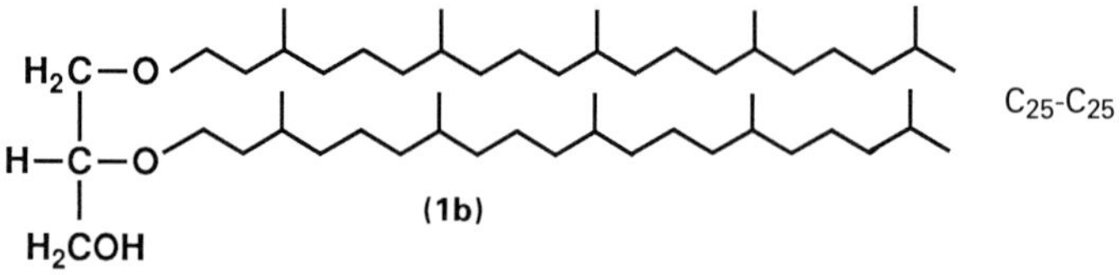

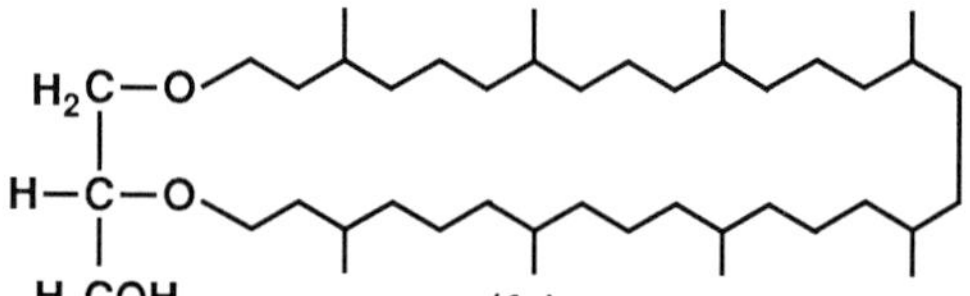

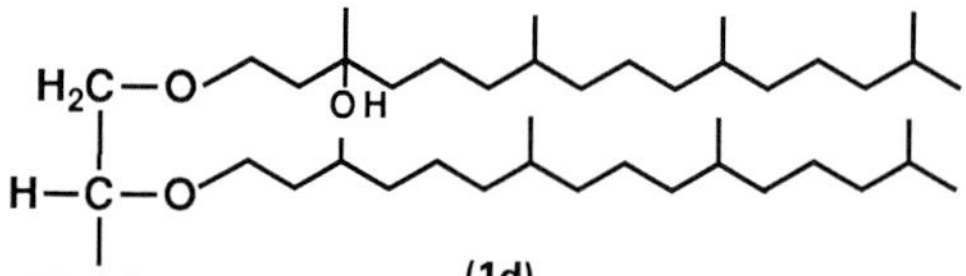

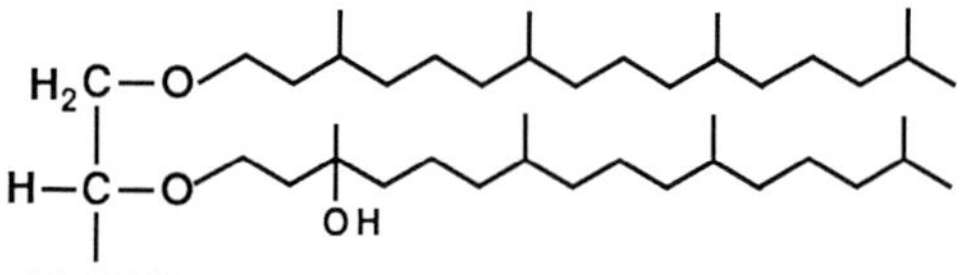

2a, Fig. 2) as well as from dibiphytanyldiglycerol tetraether or dibiphytanylglycerol nonitoltetraether containing one to four cyclopentane rings in each of the C_{40} biphytanyl groups (structures **2b**–**2i**, Fig. 2) [14].

Present nomenclature of these complex lipids is cumbersome and trivial names appear to be required. Nishihara *et al.* [15] have suggested that diphytanylglycerol diether (**1**) and its variants (**1a**–**1e**) be called 'archaeol' (modified by adding the

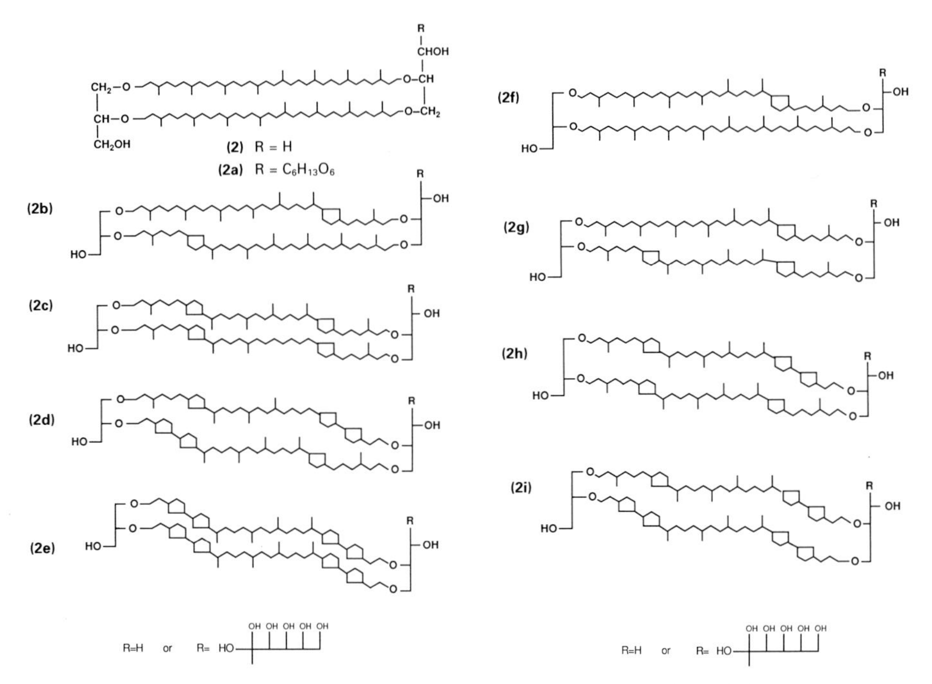

Fig. 2. Dibiphytanyldiglyceroltetraether (glycerol dialkylglycerol tetraether) (2, caldarchaeol), glycerol dialkylnonitol tetraether (2a, nonitolcaldarchaeol), and various cyclized derivatives (2b–2i, cyclized caldarchaeols and nonitolcaldarchaeols). These are the core lipids in thermoacidophiles and some methanogens.

appropriate alkyl group designations, such as C_{20}, C_{25} etc.), and dibiphytanyldiglycerol tetraether (**2**) and its variant (**2a**) be called 'caldarchaeol' and 'nonitolcaldarchaeol', respectively. This forms the basis of a useful and practical system of nomenclature for archaebacterial lipids, which will be followed here.

It should be noted that the extreme halophiles contain only archaeol-derived lipids, the methanogens contain both archaeol- and caldarchaeol-derived lipids, and the thermoacidophiles contain only cardarchaeol- and nonitolcaldarchaeol- derived lipids. Archaebacterial lipids have been the subject of several previous reviews [3,6,14,16–22].

The existence of such a large variety of unusual lipid structures in Archaea raises the question as to how these lipids are biosynthesized and how they function in the membranes of these ancient organisms. This paper will first review the structures of the polar lipids in extreme halophiles, methanogens and thermoacidophiles, and then deal with their biosynthetic pathways and their probable role in membrane structure and function.

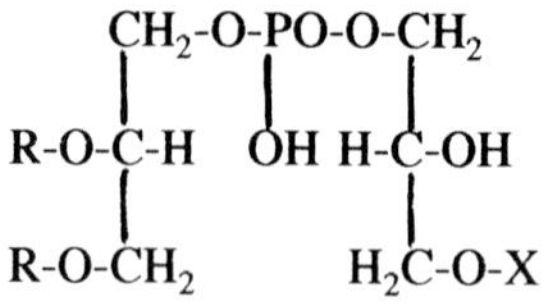

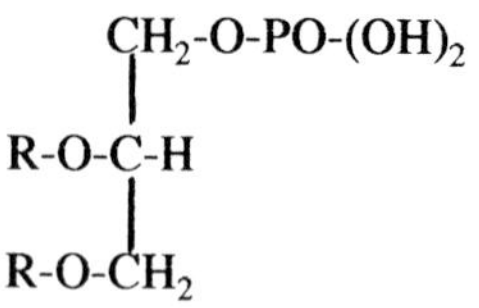

$$\begin{array}{lll} & CH_2\text{-O-PO-O-}CH_2 & \\ R\text{-O-C-H} & OH \quad H\text{-C-OH} & \\ R\text{-O-}CH_2 & \quad H_2C\text{-O-X} & \end{array}$$

$$\begin{array}{l} CH_2\text{-O-PO-}(OH)_2 \\ R\text{-O-C-H} \\ R\text{-O-}CH_2 \end{array}$$

(3) PG, X = H

(4) PGP, X = $-PO-(OH)_2$

(5) PGS, X = $-SO_2-OH$

(6) PA

R = phytanyl group $(CH_3[CH(CH_3)(CH_2)_3]_3CH(CH_3)(CH_2)_2\text{-})$

Fig. 3. Structures of archaeol phospholipids in extreme halophiles.

Polar lipids of archaebacteria

Extreme halophiles

The polar lipids of the extreme halophiles consist of one major and two or more minor phospholipids (all are acidic lipids, amino lipids being absent), and one major and two or more minor glycolipids. The phospholipid structures have been established [3,10,20] as being archaeol analogues of: phosphatidylglycerol (PG, structure **3**, Fig. 3), phosphatidylglycerol phosphate (PGP, structure **4**, Fig. 3), phosphatidylglycerol sulphate (PGS, structure **5**, Fig. 3) and phosphatidic acid (PA, structure **6**, Fig. 3) [10,23]. Stereochemically, the structures of the archaeol analogues of PG, PGP, and PGS are unusual in that both glycerol moieties have the opposite configuration to those in the corresponding diacylglycerol forms of PG and PGP found in eubacteria and eukaryotes [3].

The glycolipids include a major sulphated triglycosylarchaeol (S-TGA-1, structure **7**, Fig. 4) [3] and several minor glycolipids, such as: the desulphated TGA-1 (structure **7a**, Fig. 4); a sulphated tetraglycosylarchaeol (S-TeGA, structure **8**) and its desulphated product TeGA, structure **8a**, Fig. 4) [24]; a variant triglycosylarchaeol, TGA-2 (structure **9**, Fig. 4) [25]; and a sulphated diglycosylarchaeol, S-DGA-1 (structure **10**, Fig. 4) and its desulphated product DGA-1 (structure **10a**, Fig. 4) [26]. Examination of the structures of the glycolipids (**7**–**10**) shows that they may be derived from a basic diglycosylarchaeol, mannosylglycerolarchaeol (DGA-1, **10a**) by substitution of sugar or sulphate groups at the 3 or 6 positions of the mannose residue (see Fig. 4) [27]. However, such a structural relationship between the known halophilic glycolipids may not hold for newly discovered glycolipids of unidentified structure (e.g. DGA-2 in *Haloarcula* [25], and for the 'S-DGA-2' of a halophile strain from Japan [20]) and does not hold at all for the recently identified S-DGA-3 (structure **11**) (Fig. 5) of *Halobacterium sodomense* [28].

Recent studies [26–30] have shown that the lipid (particularly the glycolipid) composition of extreme halophiles appears to be correlated with their taxonomic

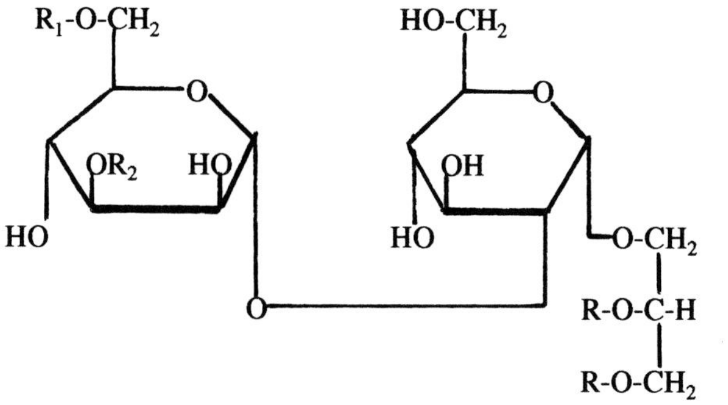

(7)	S-TGA-1, R_1 = 3-SO_3-β-Gal*p*	R_2 = H
(7a)	TGA-1, R_1 = β-Gal*p*	R_2 = H
(8)	S-TeGA, R_1 = 3-SO_3-β-Gal*p*	R_2 = α-Gal*f*
(8a)	TeGA, R_1 = β-Gal*p*	R_2 = α-Gal*f*
(9)	TGA-2, R_1 = β-Gal*p*	R_2 = H
(10)	S-DGA-1, R_1 = $-SO_2$-OH	R_2 = H
(10a)	DGA-1, R_1 = H	R_2 = H

R = phytanyl group ($CH_3[CH(CH_3)(CH_2)_3]_3CH(CH_3)(CH_2)_2^-$)

Fig. 4. Structures of archaeol glycolipids in extreme halophiles.

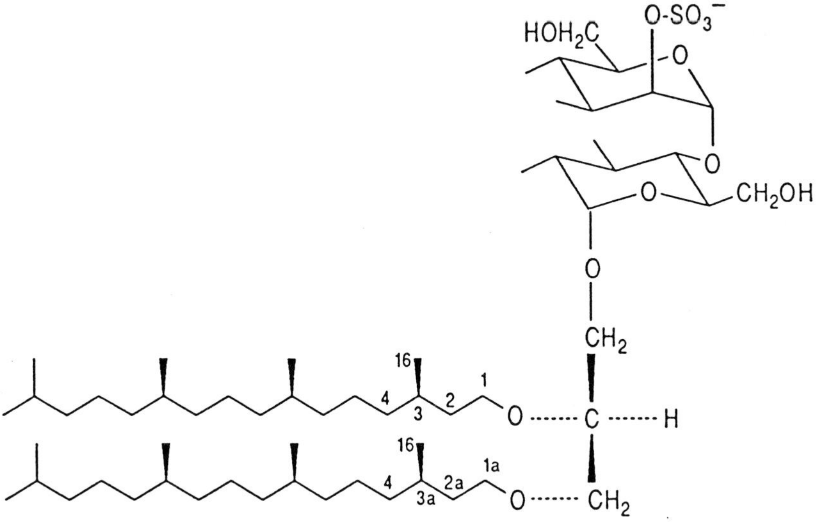

Fig. 5. Structure of the archaeol glycolipid (S-DGA-3) in *Halobacterium sodomense*: 2,3-di-O-phytanyl-1-[Man*p*-(2-sulphate)-α-(1-4)-Glc*p*-]*sn*-glycerol (11) [28].

Table 1. Distribution of polar lipids in known genera of extreme halophiles. For structures of lipids see Figs. 3 and 4.

Genus	PGP **(3)**	PG **(4)**	PGS **(5)**	PA* **(6)**	S-TGA-1 **(7)**	TGA-1 **(7a)**	S-TeGA **(8)**	TGA-2 **(9)**	S-DGA-1 **(10)**	DGA-1 **(10a)**
Halobacterium[a]	+++	+	+	tr	++	+	+	−	−	−
Haloarcula[b]	+++	+	++	tr	−	−	−	++	−	−
Haloferax[c]	++	+++	−	tr	−	−	−	−	++	+
Halococcus[d,e]	+++[h]	+[h]	−	+[e]	+[d]	tr[d]	−	−	+[e]	−
Natronobacterium[f]	+++[h]	+[h]	−	tr	−	−	−	−	−	−
Natronococcus[g]	+++[h]	+[h]	−	tr	−	−	−	−	−	−

Type species: [a] *H. cutirubrum, H. halobium, H. salinarium, H. saccharovorum* [27–30]; [b] *Ha. marismortui, Ha. valismortis, Ha. hispanica, Ha. californiae, Ha. sinaiiensis* [25,27,29]; [c] *Hf. mediterranei, Hf. volcanii, Hf. gibbonsii* [27,29]; [d] *Hc. morrhuae* [31], *Sarcina literalis* and a *Sarcina* spp. [32]; [e] *Hc. saccharolyticus* [10] contains archaeol PA, S-DGD-1 and several unidentified glycolipids including a phosphoglycolipid. [f] *Nb. pharoanis, Nb. magadii, Nb. gregoryi* and [g] *N. occultis* [33,34], also contain a cyclic form of PGP. [h] These lipids occur as both C_{20}–C_{20} and C_{20}–C_{25} species.

* First reported in a halophilic archaebacterium from a salt mine [23].

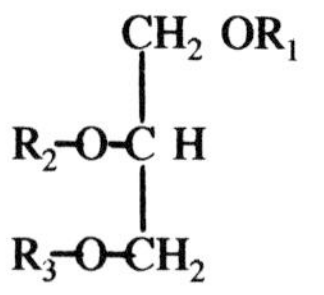

$$\begin{array}{l} \quad\ CH_2\ OR_1 \\ \quad\ \ | \\ R_2{-}O{-}C\ H \\ \quad\ \ | \\ R_3{-}O{-}CH_2 \end{array}$$

	R_1	R_2	R_3
(12)	PI, *P-myo*-inositol	C_{20}	C_{20}
(12a)	PI-A_{OH}, *P-myo*-inositol	3-OH-C_{20}	C_{20}
(13)	PE, *P*-ethanolamine	C_{20}	C_{20}
(13a)	PE-Acyc, *P*-ethanolamine	C_{20} - cyclic -	C_{20}
(13b)	PE-A_{OH}, *P*-ethanolamine	C_{20}	3-OH-C_{20}
(14)	PS, *P*-serine	C_{20}	C_{20}
(14a)	PS-A_{OH}, *P*-serine	3-OH-C_{20}	C_{20}
(15)	PGA, P-1-βGlcN.Ac	C_{20}	C_{20}
(16)	PPDA, *P*-pentanetetrol-N$(CH_3)_2$	C_{20}	C_{20}
(17)	PPTA, *P*-pentanetetrol-N$(CH_3)_3$	C_{20}	C_{20}
(18)	MGA, Glc*p*β1 - 1	C_{20}	C_{20}
(18a)	MGAcyc, Glc*p*β1 - 1	C_{20} - cyclic -	C_{20}
(18b)	MGAcyc-PE, 6-*P*-ethanolamine-Glc*p*β1 - 1	C_{20} - cyclic -	C_{20}
(19)	DGA-1, Glc*p*α 1 - 2 Gal*f*β1 - 1	C_{20}	C_{20}
(20)	DGA-2, Gal*f*β1 - 6 Gal*f*β1 - 1	C_{20}	C_{20}
(21)	DGA-3, Glc*p*β1 - 6 Glc*p*β1 - 1	C_{20}	C_{20}
(21a)	DGAcyc, Glc*p*β1 - 1	C_{20} - cyclic -	C_{20}
(22)	DGA-4, Man*p*α 1 - 3Gal*p*β1 - 1	C_{20}	C_{20}
(23)	DGA$_{OH}$, Gal*p*β1 - 6 Gal*p*β1 - 1	C_{20}	3-OH-C_{20}

Fig. 6. Complex archaeol lipids in methanogens.

classification on the level of the genera so far distinguished: *Halobacterium*, *Haloarcula*, *Haloferax*, *Halococcus*, *Natronobacterium* and *Natronococcus* (Table 1). All of these genera were found to contain archaeol PGP (**3**) as the major phospholipid, archaeol PG (**4**) and PGS (**5**) as minor phospholipids, and small to trace amounts of archaeol PA (**6**) [10,23] (see Table 1), with the following exception: PGS is characteristically absent in *Haloferax*, *Halococci*, *Natronobacteria* and *Natronococci* species. Both C_{20}–C_{20} and C_{20}–C_{25} species of PGP and PG occur in the Haloalkaliphiles [33,34] and in *H. saccharolyticus* (a Halococcus) [10]; and a cyclic form of PGP was found in *Natronococcus occultis* [34].

The glycolipid composition is more discriminating (see Table 1 and Fig. 4): (1) the *Halobacteria* contain S-TGA-1 (**7**), S-TeGA (**8**), and TGA-1 (**7a**) as major glycolipids; (2) the *Haloarcula* contain major amounts of TGA-2 (**9**) and minor amounts of an unidentified diglycosylarchaeol, DGA-2 (containing glucose and mannose); (3) the *Haloferax* contain S-DGA-1 (**10**) as their major glycolipid and DGA-1 (**10a**) as a minor glycolipid; (4) the *Halococci* contain S-TGA-1 (**7**), TGA-1 (**7a**), S-TeGA (**8**), S-DGA-1 (**10**) and an unidentified phosphoglycolipid [10]; and (5)

$$
\begin{array}{l}
\qquad\qquad\qquad\qquad\quad CH_2OR_2 \\
H_2C\text{-}O\text{-}(C_{40}H_{80})\text{-}O\text{-}C\text{-}H \\
H\text{-}C\text{-}O\text{-}(C_{40}H_{80})\text{-}O\text{-}CH_2 \\
H_2C\text{-}OR_1
\end{array}
$$

	R_1	R_2
(24)	DGC-1, Glc*p*α1-2Gal*f*β1-1	H
(24a)	PGC-1, Glc*p*α1-2Gal*f*β1-1	*P*-*sn*-3-glycerol
(25)	DGC-2, Gal*f*β1-6Gal*f*β1-1	H
(25a)	PGC-2, Gal*f*β1-6Gal*f*β1-1	*P*-*sn*-3-glycerol
(26)	DGC-3, Glc*p*β1-6Glc*p*β1-1	H
(26a)	PGC-3, Glc*p*β1-6Glc*p*β1-1	*P*-ethanolamine
(27)	CPE, H	*P*-ethanolamine
(28)	CPI, H	*P*-*myo*-inositol
(28a)	PGC-4, Glc*p*β1-6Glc*p*β1-1	*P*-*myo*-inositol
(29)	CPS, H	*P*-serine
(29a)	PGC-5, Glc*p*β1-6Glc*p*β1-1	*P*-serine

Fig. 7. Complex caldarchaeol lipids in methanogens.

neither the *Natronobacteria* nor the *Natronococci* examined contain any glycolipid components. The glycolipid composition (or absence of glycolipids) together with the phospholipid composition thus appear to be sufficient to delineate each of the known genera of extremely halophilic bacteria (Table 1) and perhaps of new genera. Examples of the latter are *Hb. sodomense* which contains a new glycolipid, S-DGA-3 (**11**, Fig. 5) [28] not found in any other known halobacterial genus, and *Halobacterium saccharovorum* which contains S-DGA-1 (**10**) as the sole glycolipid [30], suggesting that these organisms should be reclassified into new genera [21,28–30].

Methanogens

Of the 10 genera of anaerobic methanogens known so far, the following have been examined for their lipid composition: *Methanobacterium*, *Methanococcus*, *Methanospirillum*, *Methanosarcina*, and *Methanosaeta* (*Methanothrix*). The complex lipids of methanogenic bacteria are derived from caldarchaeol (**2**) and archaeol (**1**). The archaeol derivatives (Fig. 6) consist of: (*a*) phospholipids [e.g., archaeol analogues of phosphatidylinositol (PI, structure **12**), phosphatidylethanolamine (PE, structure **13**), phosphatidylserine (PS, structure **14**)] and (*b*) glycolipids (mono- and diglycosylarchaeols). The caldarchaeol lipids (Fig. 7) consist of: (*a*) phospholipids with *P*-ethanolamine, *P*-inositol or *P*-serine attached to one hydroxyl; (**b**) glycolipids with sugar groups attached to one of the hydroxyl groups; and (**c**) phosphoglycolipids with sugar groups attached to one hydroxyl and a *P*-glycerol, *P*-inositol, *P*-serine or *P*-ethanolamine group linked to the other hydroxyl.

The lipids of *Methanospirillum hungatei* [35–37] are derived from both archaeol and caldarchaeol and consist largely of two sets of phosphodiglycosylcaldarchaeols (PGCs), diglycosylcaldarchaeols, and diglycosylarchaeols, each set containing one of the following diglycosyl groups: Group 1, Glc*p*α1-2Gal*f*β or Group 2, Gal*f*β1-6Gal*f*β (Figs. 6 and 7). The phosphoglycolipids, PGC-1 and PGC-2 (structures **24a** and **25a**, Fig. 7) have the diglycosyl-1 or -2, respectively, attached to one hydroxyl of caldarchaeol and a *sn*-3-phosphoglycerol attached to the second hydroxyl; the diglycosylcaldarchaeols, DGC-1 and DGC-2 (structures **24** and **25**, Fig. 7) have the diglycosyl-1 or -2, respectively, attached to one hydroxyl of caldarchaeol; and the diglycosylarchaeols, DGA-1 and DGA-2 (structures **19** and **20**, Fig. 6) have the diglycosyl-1 or -2 attached to archaeol [35]. In addition, minor amounts of C_{20}-C_{20}-archaeol-derived phospholipids are also present: phosphatidyl-*N*,*N*-dimethylamino- (PPDA, structure **16**, Fig. 6) and *N*,*N*,*N*-trimethylaminopentanetetrol (PPTA, structure **17**, Fig. 6) [37]. The reported presence of the archaeol analogue of phosphatidyl-*sn*-1-glycerol [35] has not been confirmed [37].

Studies of the lipids of four families of methanogens [38,39] have identified the archaeol analogues of PI (**12**), PE (**13**, Fig. 6) [40] and PS (**14**, Fig. 6) as major components in Methanobacteriaceae but not in Methanomicrobiaceae and Methanosarcinaceae. Another phospholipid, caldarchaeolphosphoethanolamine (CPE, structure **27**, Fig. 7) was found only in the genera *Methanobacterium* and *Methanosarcina*. Two glycolipids were found in *Methanobacterium thermoautotrophicum*: a diglycosylarchaeol, Glc*p*β1-6Glc*p*β1-1-archaeol (DGA-3, structure **21**, Fig. 6) and the corresponding diglycosylcaldarchaeol (DGC-3, structure **26**, Fig. 7). The major lipids in the Methanobacteriaceae were the phosphoglycolipids [39,41]: diglucosylcaldarchaeolphosphoethanolamine (PGC-3, structure **26a**, Fig. 7), diglucosylcaldarchaeolphosphoinositol (PGC-4, structure **28a**, Fig. 7) and diglucosylcaldarchaeolphosphoserine (PGC-5, structure **29a**, Fig. 7). The corresponding caldarchaeolphosphoethanolamine (**27**), -phosphoinositol (**28**) and -phosphoserine (**29**, Fig. 7) were also present.

The lipids of one species of *Methanococcus*, *Mco. voltae*, are derivatives of only archaeol [42]. The major glycolipid was identified as the diglycoslyarchaeol (DGA-3, structure **21**) found also in *Mb. thermoautotrophicum* [38]; minor glycolipids were Glc*p*β1-1-archaeol (MGA, structure **18**, Fig. 6), and a novel phosphoglycosylarchaeol (PGA), GlcNAc-1-*P*-archaeol (structure **15**, Fig. 6). Lipids of the thermophilic deep-sea methanogen, *Methanococcus jannaschii* [43], are based largely on the macrocylic diether (**1c**) and consist of mono- and diglucosyl-*cyc*-archaeol (**18a** and **21a**, respectively, Fig. 6), 6-*P*-ethanolamineglucosyl-*cyc*-archaeol (**18b**) and *cyc*-archaeol-*P*-ethanolamine (**13a**); a small amount of archaeol-*P*-ethanolamine (**13**) is also present.

Ferrante *et al.* [44] have shown that the lipids of *Methanothrix* (now *Methanosaeta*) *concilii* are derived from both archaeol and hydroxyarchaeol which has a hydroxyl group at C-3 of the phytanyl chain on the *sn*-3-position (structure **1d**, Fig. 1) lipid cores. The major components are a diglycosylarchaeol (DGA-4, structure **22**, Fig. 6), a diglycosyl hydroxyarchaeol (DGA_{OH}-5, structure **23**, Fig. 6), archaeolphosphoinositol (structure **12**, Fig. 6) and hydroxyarchaeolphosphoethanolamine (structure **13b**, Fig. 6) and archaeolphosphoethanolamine (structure **13**, Fig. 6). Subsequently, Sprott *et al.* [12] showed that the lipids of *Methanosarcina barkeri* and *Methanosarcina mazei* are derived from two lipid cores: archaeol and an isomeric hydroxyarchaeol (structure **1e**, Fig. 1) in which the hydroxyl group is located at C-3 in the phytanyl chain on the *sn*-2-position of glycerol. The presence of this isomeric hydroxyarchaeol lipid core in *M. barkeri* has been confirmed by Nishihara & Koga [45] who also identified the major hydroxyarchaeol lipids as hydroxyarchaeolphosphoinositol (structure **12a**, Fig. 6) and hydroxyarchaeolphosphoserine (structure **14a**, Fig. 6). The data available so far thus indicate that several methanogenic genera have distinctive polar lipid patterns [35–39, 41–45] which might be taxonomically useful. The large variety of core lipids reported previously to be present in *M. barkeri* by De Rosa *et al.* [46], have not been confirmed by the studies of Sprott *et al.* [12] and Nishihara & Koga [45], who did not detect any C_{20}–C_{20}-tetritol diether, C_{20}–C_{25} archaeol or caldarchaeol with 1–3 cyclopentane rings.

Thermoacidophiles

The taxonomical classification of thermoacidophiles is not as complex as that of the methanogens: eight genera are known so far, distributed among four orders, Thermoplasmatales, Sulfolobales, Thermoproteales and Thermococcales. Only a relatively small number of thermoacidophiles have been examined for their lipids, mainly *Thermoplasma*, *Sulfolobus*, *Thermoproteus* and *Desulfurococcus* species, but their core lipids are more complex and varied that any of the other archaebacteria [6,14,22] (see Fig. 8), consisting of caldarchaeol (**2**) or nonitolcaldarchaeol (**2a**) with zero to four cyclopentane rings per C_{40} biphytanyl chain (**2b–2i**; Fig. 2).

The one species of *Thermoplasma* that has been examined for lipids is *Thermoplasma acidophilum*, a wall-less mycoplasma-like aerobe. Its polar lipids contain at least six different glycolipids and at least seven phosphorus-containing lipids, all based on caldarchaeol with different degrees of cyclization [6]. The major component (80 % of the polar lipids), is a phosphoglycolipid (PGC, structure **30a**, Fig. 8) based on caldarchaeol, having a single unidentified sugar attached to one hydroxyl of the tetraether and a *sn*-3-glycerophosphate group attached to the other [6,18]. The corresponding monoglycosylcaldarchaeol containing the unidentified sugar (structure **30**, Fig. 8) as well as monoglucosyl and diglucosylcaldarachaeol derivatives of uncharacterized structure (**31** and **32**, respectively) are also present. The rest of the polar lipids are mostly derived from caldarchaeol, but a few are derived from archaeol [6]. Detailed structural studies on the glycolipids of *Thermoplasma* have still to be done.

Species of the sulphur-dependent *Sulfolobus* genus live in sulphataric hot springs at temperatures between 60 °C and 100 °C and pHs between 1 and 5, and are sulphur and iron-oxidizing. The polar lipids of only three species, *Sulfolobus acidocaldarius*, *Sulfolobus solfataricus* (formerly *Caldariella acidophila*) and *Sulfolobus brierleyi* (also called 'ferrolobus') have been investigated in detail [9,14,18,19,47]. These lipids are derived almost exclusively from either caldarchaeol (**2**) or nonitolcaldarchaeol (**2a**). Both types of tetraether may also contain from one to four cyclopentane rings in their biphytanyl chains (Fig. 8), the extent of cyclization increasing with increasing growth temperature [14,48].

The polar lipids of *S. acidocaldarius*, grown either heterotrophically or autotrophically, and 'ferrolobus', a strict autotroph, are predominantly glycolipids (35 %) and phosphoglycolipids (45 %) [49,50]. The major glycolipids in *S. acidocaldarius* and 'ferrolobus' were tentatively identified as glycolipid DGC-4 (structure **32**, Fig. 8), a glucosylgalactosylcaldarchaeol, and glycolipid GNC (structure **39**, Fig. 8), a glucosylnonitolcaldarchaeol, the glucosyl group being linked to one of the hydroxyl groups of the nonitol group [50].

$$\begin{array}{l} \qquad\qquad\qquad CH_2\text{-}O\text{-}R_2 \\ H_2C\text{-}O\text{-}(C_{40}H_{80})\text{-}O\text{-}C\text{-}H \\ H\text{-}C\text{-}O\text{-}(C_{40}H_{80})\text{-}O\text{-}CH_2 \\ R_1\text{-}O\text{-}CH_2 \end{array}$$

		R_1	R_2	No. of cyclopentane rings
(30)	MGC,	monoglycosyl	H	0–2
(30a)	PGC,	monoglycosyl	*P*-*sn*-3-glycerol	0–2
(31)	MGC-1,	Glc*pβ*	H	0–4
(31a)	PGC-6,	Glc*pβ*	*P*-inositol	0–2
(32)	DGC-4,	Glc*pβ*- Gal*pβ*	H	0
(32a)	PGC-7,	Glc*pβ*- Gal*pβ*	*P*-inositol	0
(33)	DGC-5,	Glc*pβ*1 - 4 Gal*pβ*	H	0
(33a)	PGC-8,	Glc*pβ*1 - 4 Gal*pβ*	*P*-inositol	0
(34)	PGC-9,	Gal*pβ*	*P*-inositol	0
(35)	DGC-3,	Glc*pβ*1 - 6 Glc*pβ*	H	0–4
(36)	TGC,	(Glc*pβ*1 - 6)$_2$ Glc*pβ*	H	0–4
(36a)	PGC-10,	(Glc*pβ*1 - 6)$_2$ Glc*pβ*	*P*-inositol	0–4
(37)	PGC-11,	Gal*pβ*	*P*-inositol	0
(38)	DGC-6,	Glc*pα* 1 - 4 Gal*pβ*	H	0
(38a)	PGC-12,	Glc*pα*1 - 4 Gal*pβ*	*P*-inositol	0

$$\begin{array}{l} \qquad\qquad\qquad\qquad\qquad\qquad\qquad\quad CH_2OR_2 \\ \qquad\qquad\quad H_2C\text{-}O\text{-}(C_{40}H_{80})\text{-}O\text{-}C\text{-}H \\ \qquad\qquad\quad H\text{-}C\text{-}O\text{-}(C_{40}H_{80})\text{-}O\text{-}CH_2 \\ \qquad\qquad\quad HO\text{-}C\text{-}H \\ HO\text{—}CH_2\text{-}CH\text{—}CH\text{—}CH\text{—}C\text{—}CH_2OH \\ \qquad\quad OH \quad OR_1 \quad OH \quad OH \end{array}$$

(39)	GNC,	Glc*pβ*	H	0–4
(39a)	PGNC,	Glc*pβ*	*P*-inositol	0–4
(39b)	GNC,	HSO3Glc*pβ*	H	0–4

Fig. 8. Complex caldarchaeol and nonitolcaldarchaeol lipids of thermoacidophiles.

The phosphoglycolipids in both *S. acidocaldarius* and 'ferrolobus' [48] consisted predominantly (72%) of the phosphoinositol derivatives of glycolipids DGC-4 (**32**) and GNC (**39**): Glc-Gal-caldarchaeol-*P*-inositol (PGC-7; structure **32a**, Fig. 8), and Glc-nonitolcaldarchaeol-*P*-inositol (PGNC, structure **39a**, Fig. 8), in which the glucosyl group is linked to one hydroxyl group in the nonitol group and the phosphoinositol is linked to the free hydroxyl group in the glycerol residue of caldarchaeol. A minor acidic glycolipid detected in *S. acidocaldarius* cells, but not in 'ferrolobus', was partially characterized as a monosulphate derivative of glycolipid GNC (structure **39b**, Fig. 8); the sulphate group is presumably attached to the glucosyl group but its precise location is not known [19]. The main glycolipids of *S. acidocaldarius* and 'ferrolobus' are probably similar, if not identical, to those in *S. solfataricus* (see below), but further structural studies are necessary to establish this unambiguously [6,14].

The glycolipids of *S. solfataricus* [14,47,51] have been characterized in more detail than those of *S. acidocaldarius*, but their structures are still not unambiguously defined. The major glycolipids are the caldarchaeol glycolipid DGC-4 (6%), (structure **32**, Fig. 8) and the nonitolcaldarchaeol glycolipid GNC (14%), (structure **39**, Fig. 8); the linkage positions of the disaccharide in (**32**) and the position of attachment of the glucosyl group to the nonitol group in (**39**) have still to be determined. The phosphoglycolipids consisted mainly of the phosphoinositol derivatives of glycolipids (**32**) (DGC-4) and (**39**) (GNC): Glc*pβ*-Gal*pβ*-caldarchaeol-*P*-inositol (8%, PGC-7 structure **32a**, Fig. 8), and Glc*pβ*-nonitolcardarchaeol-*P*-inositol (55%, PGNC, structure **39a**, Fig. 8) respectively; the sulphated derivative of glycolipid (**39**), HSO_3-Glc*pβ*-nonitolcaldarchaeol (10%, structure **39b**, Fig. 8) was also identified.

One species of the anaerobic, sulphur-dependent *Thermoproteus* genus, *Thermoproteus tenax*, was found to contain 5% neutral lipids, 20% glycolipids and 75% phospholipids and phosphoglycolipids [52]. The ether core lipids consisted mainly (95%) of caldarchaeols with minor proportions (5%) of archaeols. The caldarchaeols are acyclic (**2**), or contain one (**2b**), or two (**2c**) cyclopentane rings in each chain, as well as mixed numbers of rings (structures **2f**, **2g**, **2h**, Fig. 2), similar to those found in *Sulfolobus* [14]. The polar lipid composition resembles that of *Sulfolobus* but is characteristically different with respect to the glycosyl groups of the glycolipids. The glycolipids of *T. tenax* consist of mono-, di-, and triglucosyl-caldarchaeols, the sugar groups being *β*1-6-linked (structures **31**, **35** and **36**, Fig. 8, respectively) and the caldarchaeol cores contain from zero to four rings (Fig. 8). It is of interest that, except for the presence of rings in the caldarchaeol core, the diglucosylcaldarchaeol (**35**) is identical to that found in the methanogen *Mb. thermoautotrophicum* (structure **26**, Fig. 7) and the monoglucosylcaldarchaeol (structure **31**, Fig. 8) is similar to (**30**) in *Thermoplasma*, but the triglucosylcaldarchaeol (structure **36**, Fig. 8) has not been reported in any other archaebacterium [52]. The major phosphoglycolipids are the monoglucosylcaldarchaeol with phosphoinositol attached to the free hydroxyl group (PGC-6, structure **31a**, Fig. 8), which is a variation of structure (**30a**) in *Thermoplasma*, and the phosphoinositol derivative (PGC-10, structure **36a**, Fig. 8) of the triglucosylcaldarchaeol (**36**). No nonitolcaldarchaeol glyco- or phosphoglycolipids appear to be present in *T. tenax* [52].

Thermococcales

The lipids of one species of this anaerobic sulphur-reducing group of thermoacidophiles, *Desulfurococcus mobile*, are derived from caldarchaeol and resemble those in *Sulfolobus*, except that the lipid cores do not contain any nonitolcaldarchaeols. The major glycolipid is Glc*pα*1-4Gal*pβ*-caldarchaeol (structure **38**, Fig. 8), and the phosphoglycolipids are Gal*pβ*-cardarchaeol-*P*-inositol (PGC-11, structure **37**, Fig. 8) and Glc*pα*1-4Gal*pβ*-cardarchaeol-*P*-inositol (PGC-12, structure **38a**, Fig. 8) [53].

Biosynthetic pathways

Extreme halophiles

Biosynthesis of the archaeol analogues of phospholipids and glycolipids proceeds by complex pathways in a multienzyme, membrane-bound system that is absolutely dependent on 4 M-NaCl concentration [16,20,21]. Synthesis of the isoprenoid/isopranoid chains in the archaeol lipid core takes place via the mevalonate pathway for isoprenoids, also absolutely dependent on 4 M-NaCl concentration [16], starting from acetate (and involving lysine [54]), and proceeding to mevalonate, isopentenyl-*PP* and then to geranyl-*PP*, farnesyl-*PP* and geranylgeranyl-*PP* [16,20] (Fig. 9).

Final reduction of geranylgeranyl-*PP* to the phytanyl group takes place only after linkage to the glycerol backbone, to form an unidentified C_{20}-isoprenyl glycerol ether derivative ('pre-diether') which is the precursor of both phospholipids and glycolipids (see Figs. 9 and 10). The 'pre-diether' may be formed by an unknown mechanism whereby a suitably substituted glycerol derivative, tentatively dihydroxyacetone (DHA), is alkylated with a C_{20}-isoprenyl-*PP* (geranylgeranyl-*PP*) [20,55]. The choice of DHA, or a substituted DHA, as acceptor of the isoprenyl group is based on the finding that the glycerol moiety of diphytanylglycerol undergoes dehydrogenation at C-2 but not at C-1 (or C-3) [20], thus eliminating triose phosphates (DHA-*P* and glyceraldehyde-*P*) and glycerol-*P* as possible acceptors, since these would exchange hydrogen at C-1 by aldo–keto or keto–enol isomerizations (see Fig. 9). Glycerol alone (or a substituted glycerol) might also serve as an acceptor, as has been suggested previously [14], but it would have to undergo dehydrogenation at C-2 at some stage to account for the observed loss of hydrogen at C-2 in phytanylglycerol lipids of *Halobacterium cutirubrum* [20].

Evidence supporting the involvement of an isoprenyl-*PP* as donor of the C_{20}-isoprenyl group is provided by the demonstration that bacitracin, which complexes with polyprenylpyrophosphates, is a powerful inhibitor of the biosynthesis of the 'pre-diether' and of phospholipids and glycolipids in whole cells of *H. cutirubrum* [56].

Pulse-labelling studies with whole cells of *H. cutirubrum* using [^{32}P]phosphate or [^{14}C]glycerol as precursors showed that the phospholipids were labelled in the order pre-PA > phospholipid precursor > pre-PGP > pre-PG > pre-PGS [55]. These results suggest the following pathway for the biosynthesis of phospholipids in extreme halophiles (Figs. 9 and 10): the pre-diether is phosphorylated with ATP, and the pre-PA formed is then converted into the 'phospholipid precursor', most likely CDP-archaeol, which is then reacted with glycerophosphate to form pre-PGP; dephosphorylation of pre-PGP by a specific phosphatase would give rise to pre-PG, which could then be converted into pre-PGS by reaction with 3′-phosphoadenosine-5′-phosphosulphate (PAPS). All of the 'pre-' phospholipids are subsequently hydrogenated to give the final, saturated archaeol analogues of PGP, PG and PGS.

The pre-diether also serves as the precursor of the glycolipids [55] (Figs. 9 and 10), being glycosylated stepwise with glucose, mannose and galactose followed by sulphation with PAPS to give the pre-S-TGA which is finally reduced to the saturated S-TGA (**7**). Evidence in support of a stepwise glycosylation and sulphation has been provided by pulse-labelling studies with whole cells of *H. cutirubrum* grown

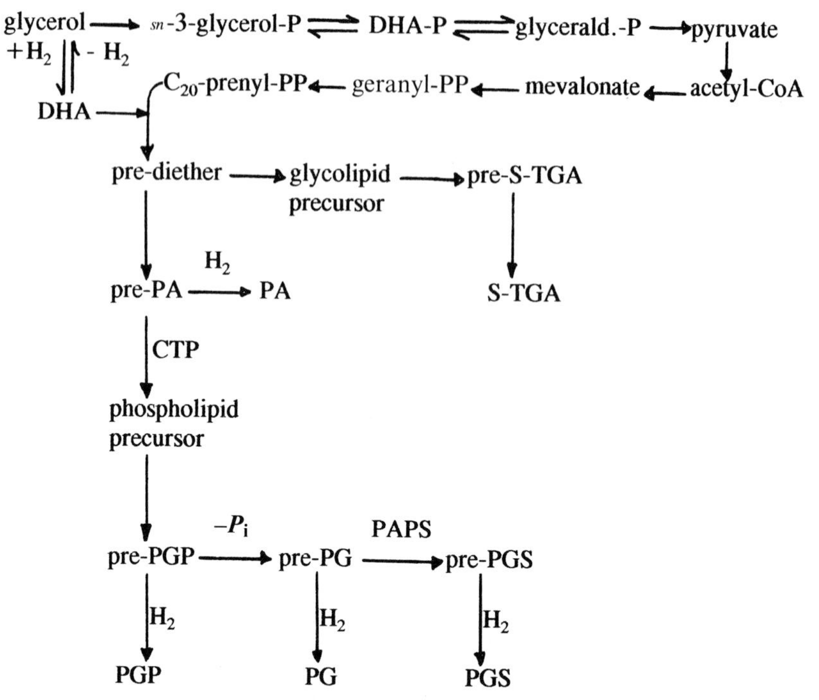

Fig. 9. Proposed pathways for biosynthesis of archaeol phospholipids and glycolipids in extreme halophiles.

in presence of [^{35}S]sulphate or [^{14}C]glycerol, which established the following product–precursor relationship between the glycolipids of this bacterium (P. Deroo & M. Kates, unpublished work) (see Fig. 4):

$$\text{glycolipid precursor} \xrightarrow{\text{UDP-Glc}} \text{Glc-archaeol (MGA-1)} \xrightarrow{\text{UDP-Man}}$$
$$\text{DGA-1} \xrightarrow{\text{UDP-Gal}} \text{TGA-1} \xrightarrow{\text{PAPS}} \text{S-TGA-1.}$$

The minor glycolipid S-TeGA (**8**) could be biosynthesized by galactofuranosylation of S-TGA-1 (**7**) or by galactofuranosylation of TGA-1 (**7a**) followed by sulphation with PAPS. The minor non-sulphated glycolipids TGA-1 and TeGA-1 could be formed by deletion of the appropriate sulphation steps, or by the action of sulphatases on S-TGA-1- and S-TeGA. Such pathways would account for the presence of the glycolipids found in the *Halobacteria* (Table 1). In *Haloferax* species, the major glycolipid S-DGA-1 (**10**) (Fig. 4) might be formed by deletion of the galactosylation step of DGA and insertion of a specific DGA sulphation step. In *Haloarcula*, the major glycolipid TGA-2 (**9**) (Fig. 4) might be formed by replacement of the galactosylation step with a glucosylation reaction using UDP-Glc.

Thus the characteristic glycolipid composition of the three genera of halobacteria (Table 1) could be achieved by deletion and/or insertion of the appropriate genes for glycosylating enzymes and sulphating enzymes. Studies on the isolation of the enzymes and corresponding genes involved in glycolipid and

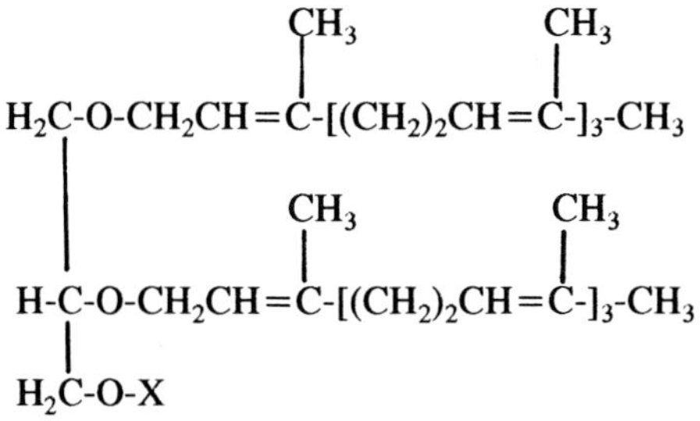

$$
\begin{array}{l}
\qquad\qquad\qquad\quad CH_3 \qquad\qquad\quad CH_3 \\
H_2C\text{-}O\text{-}CH_2CH{=}C\text{-}[(CH_2)_2CH{=}C\text{-}]_3\text{-}CH_3 \\
\quad | \qquad\qquad\qquad\quad CH_3 \qquad\qquad\quad CH_3 \\
H\text{-}C\text{-}O\text{-}CH_2CH{=}C\text{-}[(CH_2)_2CH{=}C\text{-}]_3\text{-}CH_3 \\
\quad | \\
H_2C\text{-}O\text{-}X
\end{array}
$$

pre-diether	X = unidentified
pre-PA	X = -PO-(OH)_2
pre-PG	X = $\text{-PO(OH)-O-CH}_2\text{CH(OH)CH}_2\text{-OH}$
pre-PGP	X = $\text{-PO(OH)-O-CH}_2\text{CH(OH)CH}_2\text{-O-PO(OH)}_2$
pre-PGS	X = $\text{-PO(OH)-O-CH}_2\text{CH(OH)CH}_2\text{-O-SO}_3\text{-OH}$
Phospholipid precursor	X = $\text{-}\overset{O}{\overset{\Vert}{\underset{O^-}{\underset{\vert}{P}}}}\text{-O-}\overset{O}{\overset{\Vert}{\underset{O^-}{\underset{\vert}{P}}}}\text{-O-cytidine}$
Glycolipid precurosr	X = unidentified
pre-S-TGA	X = $\text{-Glc-Man-Gal-3-SO}_3\text{-}$

Fig. 10. Structures of isoprenyl precursors of archaeol phospholipids and glycolipids of extreme halophiles.

phospholipid biosynthesis would help to establish the biosynthetic pathways of these lipids with greater certainty, and to elucidate the taxonomic relationships between the genera of extreme halophiles.

Methanogens

No detailed studies have as yet been reported on the biosynthesis of archaeol or caldarchaeol glycolipids or phosphoglycolipids in methanogens. Nevertheless, examination of the known structures of methanogen lipids (Figs. 6 and 7) suggests possible biosynthetic relationships between archaeol and caldarchaeol complex lipids that could be tested experimentally. It may be assumed first that the archaeol molecule would be synthesized by the same or similar route as in extreme halophiles (see Figs. 9 and 10). On the basis of the presence in *Methanospirillum hungatei* of analogous archaeol and caldarchaeol glyco- and phosphoglycolipids, Kushwaha *et al.*, [35] have proposed that the biosynthesis of the phosphoglycosylcaldarchaeols PGC-1 (**24a**) and

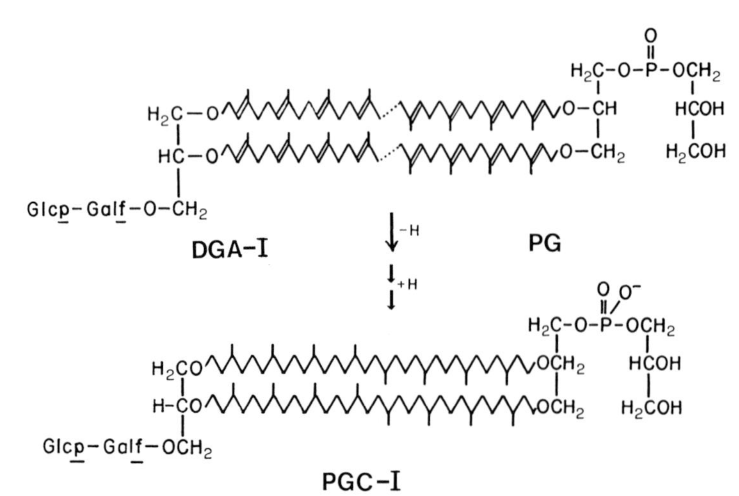

Fig. 11. Proposed mechanism for biosynthesis of caldarchaeol phosphoglycolipids in methanogens, e.g. *M. hungatei* [35].

PGC-2 (**25a**) might occur by head to head condensation of a molecule of *sn*-3-glycerophosphoarchaeol with DGA-1 (**19**) or DGA-2 (**20**), respectively. More likely, condensation would occur between the isoprenyl ether precursors followed by hydrogenation (see Fig. 11). However, it should be noted that the required *sn*-3-glycerophosphoarchaeol (or its isoprenyl ether derivative) has not been identified in *M. hungatei* [35,37].

In recent [^{32}P]phosphate pulse-chase studies with *Mb. thermoautotrophicium* [41] a pronounced lag was observed between labelling of the diether and tetraether polar lipids, consistent with the mechanism shown in Fig. 11. This mechanism would apply to the biosynthesis in *Mb. thermoautotrophicum* of the phosphoglycosylcaldarchaeols containing phosphoethanolamine, phosphoserine and phosphoinositol, since all of the corresponding archaeol phospholipids and glycolipids are present (Fig. 6) [41].

However, it should be noted that no direct experimental evidence is available concerning the mechanism of this unique head-to-head condensation (Fig. 11), other than that the two carbons involved in the coupling are derived from C-2 of mevalonate [14]. It is plausible that the head-to-head condensation would involve coupling of terminal carbons on isoprenyl intermediates, presumably isoprenyl ether archaeol derivatives similar to those detected in extreme halophiles (see Figs. 9 and 10). Recent labelling studies with *M. hungatei* have indeed eliminated the possibility of coupling of saturated archaeol termini, and favour the coupling of termini in the isoprenyl analogues [57].

Thermoacidophiles

Biosynthesis of caldarchaeol and nonitolcaldarchaeol lipids in thermoacidophiles has been studied in whole cells of *S. solfataricus* using [U-^{14}C, 1(3)-^{3}H]glycerol and [U-^{14}C, 2-^{3}H]glycerol as precursors [14,58]. It was found that the biphytanyl moieties of both caldarchaeol and nonitolcaldarcheol undergo complete loss of ^{3}H from [2-^{3}H]glycerol and 50% loss from [1(3)-^{3}H]glycerol, as expected for biosynthesis of acetate from glycerol via the glycolytic pathway, and consistent with the findings for phytanyl groups in *H. cutirubrum* [20]. The glycerol moieties of caldarchaeol and nonitolcaldarchaeol, however, do not undergo loss of hydrogen either from C-1(3) or from C-2, nor does the free glycerol of glycerophosphate pools undergo any dehydrogenation [14]. This is in contrast with the situation in extreme halophiles in which the archaeol glycerol moiety loses hydrogen at C-2 but not at C-1(3).

The glycerol precursors of caldarchaeol or nonitolcaldarchaeol thus cannot undergo dehydrogenation at C-2, nor aldo–keto or keto–enol isomerizations, thus eliminating the involvement of DHA, DHA-*P*, and glyceraldehyde-*P*. It was concluded [14,58] that the ether-forming step could occur in thermoacidophiles by direct condensation of geranylgeranyl-*PP* or similar allylic pyrophosphates with glycerol as acceptor. However, an activated or derivatized glycerol would perhaps be a more appropriate acceptor, as in the extreme halophiles.

Presumably, the products of this alkylation reaction would be isoprenyl ether intermediates similar to those detected in the biosynthesis of archaeol lipids in *H. cutirubrum* (Fig. 10) [55] and in *M. hungatei* [57]. However, in thermoacidophiles, these prenyl ether intermediates would probably not be reduced directly, but would first participate in head-to-head condensations to form the cardarchaeol polar lipids (see Fig. 8); introduction of cyclopentane rings would also occur before final reduction of the chains [22].

Biosynthesis of the nonitol group in nonitolcaldarchaeol could occur by aldol or acetoin-type condensation between a triose and a hexose precursor, followed by reduction [14,58]. Evidence favouring this mechanism is the complete loss of ^{3}H from [2-^{3}H]glycerol, 70 % retention of ^{3}H from [1(3)-^{3}H]glycerol at carbons 1–3, and the incorporation of labelled glucose or fructose into carbons 4–9 in the nonitol skeleton [14,22,58]. It is clear that detailed studies with whole-cell and cell-free systems are now required to elucidate the novel mechanisms underlying the complex biosynthetic pathways of archaebacterial lipids.

Function of archaebacterial lipids

The fact that all archaebacterial membranes contain lipids derived from diphytanylglycerol diether (**1**) or its dimer dibiphytanyldiglycerol tetraether (**2**) suggests that these ‘peculiar’ lipids have functions that are common to all archaebacteria [6,21,22]. Thus, in general, the alkyl ether structure, in contrast with the usual acyl ester structure, should impart stability to the lipids over the wide range of pH encountered by these bacteria; the saturated alkyl chains would impart stability towards oxidative degradation, particularly in the extreme halophiles that are exposed to air and sunlight; the branched isopranyl structure of archaeol and caldarchaeol would ensure that the membrane lipids are in the liquid crystalline state at ambient temperatures; and the covalent linking of the ends of the chains together with the introduction of pentacyclic rings would keep the fluidity fairly constant as the temperature increased [14,48]. Furthermore, the ‘unnatural’ *sn*-1 configuration of the backbone glycerol would impart resistance to attack by phospholipases released by other organisms and would thus have a survival value for the extreme halophiles. More specific functions of archaebacterial lipids will now be considered.

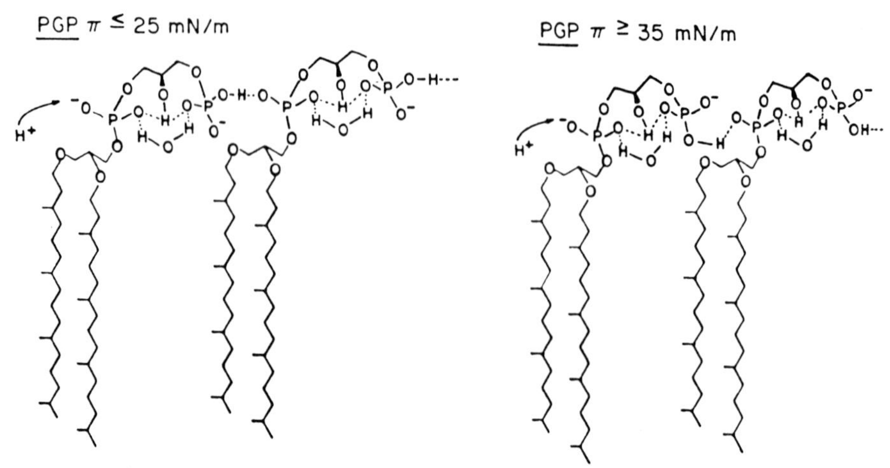

Fig. 12. A hypothetical model of a proton conductance pathway in the red membrane outer layer consisting largely of PGP. The model shows part of the PGP monolayer at a water–air interface at low (< 25 mN/m) and high (> 35 mN/m) surface pressure (reproduced from [61] with permission).

In extreme halophiles the sulphated triglycosylarchaeol (S-TGA-1, **7**) is associated exclusively with the purple membrane (PM) of *Halobacteria* [59], in which it is located entirely in the exterior surface layer of the membrane [60]. S-TGA-1, together with the major phospholipid PGP (**4**) (in mole ratio 1:1 [21]) are capable of participating in proton conductance pathways [61]. Such pathways, involving the sulphate of the polar headgroup of S-TGA and the phosphate groups of PGP would serve to transport the protons transduced by light-activated bacteriorhodopsin (BR) across the outer surface of the PM to the red membrane, where the PGP headgroup phosphates would conduct the protons to the sites of the H^+-ATPases in the red membrane, to drive ATP synthesis [62]. A hypothetical model of a proton conductance pathway involving the polar headgroup of PGP in the red membrane is shown in Fig. 12. Note the participation of the free-hydroxyl of the headgroup glycerol in the hydrogen bonding network, which has now been demonstrated by Fourier transform infra-red studies of PGP and its deoxy analogue, dPGP [63].

Another point that should be noted is the maintenance of the number of negative charges per mole of anionic lipid in the membranes of *Halobacteria*, *Haloarcula* and *Haloferax* at around 1.5–2.0, despite the differences in polar lipid composition (Table 1) [21]. Such a high negative-charge surface density would be shielded by the high Na^+ ion concentration (4 M), thus preventing disruption of the lipid bilayers due to negative-charge repulsion. A highly negatively charged membrane surface would thus appear to be required for the survival of extreme halophiles in media of high salt concentration.

The sulphated triglycosylarchaeol may have another function in the PM, related to the proton pumping action of BR, since it has been demonstrated that reconstitution of BR in PGP vesicles containing S-TGA-1 results in increased rates of proton pumping [64]. Similar functions might be envisaged for the sulphated diglycosylarchaeol (S-DGA-1, **10**) in species of *Haloferax*, which also contain the PM, but no experimental evidence is available concerning the function of this glycolipid.

As was discussed above, the membranes of caldarchaeol-containing methanogens and thermoacidophiles consist of essentially bipolar monolayer structures. These structures would impart stability to the membrane at the high growth temperatures of thermophilic methanogens and of thermoacidophiles [22]. In the methanogens, possession of both archaeol and caldarchaeol-based membrane lipids would also be an advantage under conditions of high methane concentration that must be present in these cells and might lead to membrane disruption. However, it should be noted that some species of methanogens, e.g. *M. voltae*, manage to survive with only the archaeol-type lipids in their membranes.

Another methanogen, *Methanococcus jannaschii*, isolated from a deep-sea hydrothermal vent at temperatures of 85 °C or greater, has membrane lipids based on *cyc*-archaeol (**1c**), which was not found in any other species of methanogens surveyed and may be unique to methanogens from deep-sea hydrothermal vents [11,43]. It is possible that the presence of *cyc*-archaeol-based lipids may be related to the high pressures under which these deep-sea methanogens live.

The high proportions of glycosylated caldarchaeols present in membranes of both methanogens and thermoacidophiles may further stabilize the membrane structure by interglycosyl headgroup hydrogen bonding. In addition, as was discussed above, the presence of cyclopentane rings in the biphytanyl chains of caldarchaeol may serve to fine-tune the rigidity of the membrane monolayer in direct response to the growth temperature of the thermoacidophile [14,22,48]. In contrast, there does not appear to be any specific structural feature of the polar lipids of thermoacidophiles that would be related to the low pH of their growth environment; for example, no amino or other basic groups are present that might serve to protect the membrane against the high H^+ concentration in the external environment.

Another puzzling point concerns the asymmetric orientation at the exterior membrane surface of the glycosyl groups in the caldarchaeol lipids of thermoacidophiles and some methanogens, and hence the orientation of the anionic groups (phosphates in phosphoglycocaldarchaeols) into the inner membrane layer, thus placing a highly negatively charged surface density on one side of the membrane. It is difficult to understand how the membrane would remain stable under such an arrangement, unless the negative charges on the interior side were neutralized or shielded by protonated amino groups in the membrane proteins.

Summary and conclusions

The foregoing review of membrane lipids in archaebacteria has revealed a remarkable variety of polar lipids classes, including phospholipids, glycolipids, phosphoglycolipids and sulpholipids, all derived from the one basic core structure, diphytanylglycerol (**1**) and an equally remarkable set of novel pathways for their biosynthesis. Even with the relatively limited knowledge that we have of the physical properties of these lipids, it is clear that they are well-adapted as membrane components to the particular environmental conditions of the three groups of archaebacteria, extreme halophiles, methanogens, and thermoacidophiles. However, much remains to be learned concerning the precise asymmetric arrangement of the lipids in the membrane bilayers or monolayers, the interaction of the lipids with the membrane proteins, and the function of this membrane lipid asymmetry with respect to ion transport, permeability to nutrients, proton transport and conductance, and energy transduction. Perhaps then these unusual lipids will not appear so strange and our knowledge of them will help us to understand the function of the more familiar lipids in the eubacteria and eukaryotes.

Support by the Natural Sciences and Engineering Research Council of Canada and the Medical Research Council of Canada is acknowledged.

Note added in proof

Recent FAB-mass spectrometry studies [65,66] have indicated that the terminal phosphate of archaeol PGP (**4**, Fig. 3) is monomethylated. This revised structure (PGP-Me) has now been confirmed by ^{1}H- and ^{13}C-n.m.r. spectrometry and partial hydrolysis studies (M. Kates *et al.*, unpublished work).

References

1. Woese, C. R. & Wolfe, R. S. (eds.) (1985) The Bacteria vol. 8: Archaebacteria, Academic Press, New York
2. Woese, C. R., Kandler, O. & Wheelis, M. L. (1990) Proc. Natl. Acad. Sci. U.S.A. **87**, 4576–4579
3. Kates, M. (1978) Prog. Chem. Fats Other Lipids **15**, 301–342
4. Langworthy, T. A. (1977) Biochim. Biophys. Acta **487**, 37–50
5. Kushwaha, S. C., Kates, M., Sprott, G. D. & Smith, I. C. P. (1981) Science **211**, 1163–1164
6. Langworthy, T. A. (1985) in The Bacteria: Archaebacteria: Lipids of Archaebacteria (Woese, C. R. & Wolfe, R. S., eds.), vol. 8, pp. 459–597, Academic Press, New York
7. Heathcock, C. H., Finkelstein, B. L., Aoki, T. & Poulter, C. D. (1985) Science **229**, 862–864
8. De Rosa, M., Gambacorta, A., Nicolaus, B., Ross, H. N. M., Grant, W. D. & Bu'Lock, J. D. (1982) J. Gen. Microbiol. **128**, 343–348
9. De Rosa, M., Gambacorta, A., Nicolaus, B. & Grant, W. D. (1983) J. Gen. Microbiol. **129**, 2333–2337
10. Moldoveanu, N., Kates, M., Montero, C. G. & Ventosa, A. (1990) Biochim. Biophys. Acta **1046**, 127–135
11. Comita, P. B. & Gagosian, R. B. (1983) Science **222**, 1329–1331
12. Sprott, G. D., Ekiel, I. & Dicaire, C. (1990) J. Biol. Chem. **265**, 13735–13740
13. Nishihara, M. & Koga, Y. (1991) Biochim. Biophys. Acta **1082**, 211–217
14. De Rosa, M., Gambacorta, A. & Gliozzi, A. (1986) Microbiol. Rev. **50**, 70–80
15. Nishihara, M., Morii, H. & Koga, Y. (1987) J. Biochem. **101**, 1007–1115
16. Kates, M. & Kushwaha, S. C. (1978) in Energetics and Structure of Halophilic Microorganisms: Biochemistry of the Lipids of Extremely Halophilic Bacteria. (Caplan, S. R. & Ginzburg, M., eds.), pp. 461–480, Elsevier, Amsterdam
17. Kates, M. (1988) in Biological Membranes: Aberrations in Membrane Structure and Function; Structure, Physical Properties and Function of Archaebacterial Lipids (Karnovsky, M. L., Leaf, A. & Bolis, L. C., eds.), pp. 357–384, Alan Liss, New York

18. Langworthy, T. A., Tornabene, T. G. & Holzer, G. (1982) Zentralbl. Bakteriol. Hyg. Abt. 1 Orig. C **3**, 228–244
19. Langworthy, T. A. (1982) Curr. Topics Membr. Transp. **17**, 45–77
20. Kamekura, M. & Kates, M. (1988) in Halophilic Bacteria: Lipids of Halophilic Archaebacteria (Rodriguez-Valera, F., ed.), pp. 25–54, CRC Press, Boca Raton, Florida
21. Kates, M. (1990) in Glycolipids, Phosphoglycolipids and Sulfoglycolipids: Glycophosphoglyco- and Sulfoglycolipids of Bacteria (Kates, M., ed.), pp. 1–122, Plenum, New York
22. De Rosa, M., Trincone, A., Nicolaus, B. & Gambacorta, A. (1991) in Life Under Extreme Conditions: Archaebacteria: Lipids, Membrane Structures, and Adaptation to Environmental Stresses (di Prisco, G., ed.), pp. 61–87, Springer, Berlin
23. Lanzotti, V., Nicolaus, B., Trincone, A., De Rosa, M., Grant, W. D. & Gambacorta, A. (1989) Biochim. Biophys. Acta **1002**, 398–400
24. Smallbone, B. W. & Kates, M. (1981) Biochim. Biophys. Acta **665**, 551–558
25. Evans, R. W., Kushwaha, S. C. & Kates, M. (1980) Biochim. Biophys. Acta **619**, 533–544
26. Kushwaha, S. C., Kates, M., Juez, G., Rodriguez-Valera, E. & Kushner, D. J. (1982) Biochim. Biophys. Acta **711**, 19–25
27. Kushwaha, S. C., Juez-Perez, G., Rodriguez-Valera, E., Kates, M. & Kushner, D. J. (1982) Can. J. Microbiol. **28**, 1365–1372
28. Trincone, A., Nicolaus, B., Lama, L., De Rosa, M., Gambocorta, A. & Grant, W. D. (1990) J. Gen. Microbiol. **136**, 2327–2331
29. Torreblanca, M. F., Rodriguez-Valera, F., Juez, G., Ventossa, A., Kamekura, M. & Kates, M. (1986) Syst. Appl. Microbiol. **8**, 89–99
30. Lanzotti, V., Nicolaus, B., Trincone, A. & Grant, W. D. (1988) FEMS Microbiol. Lett. **55**, 223–228
31. Kocur, M. & Hodgkiss, W. (1973) Int. J. Syst. Bacteriol. **23**, 151–156
32. Kates, M., Palameta, B., Joo, C. N., Kushner, D. J. & Gibbons, N. E. (1966) Biochemistry **5**, 4092–4099
33. De Rosa, T., Gambacorta, A., Grant, W. D., Lanzotti, V. D. & Nicolaus, B. (1988) J. Gen. Microbiol. **134**, 205–211
34. Lanzotti, V., Trincone, A., De Rosa, M., Grant, W. D. & Gambacorta, A. (1989) Biochim. Biophys. Acta **1001**, 31–34
35. Kushwaha, S. C., Kates, M., Sprott, G. D. & Smith, I. C. P. (1981) Biochim. Biophys. Acta **664**, 156–173
36. Grant, W. D., Pinch, G., Harris, J. E., De Rosa, M. & Gambacorta, A. (1985) J. Gen. Microbiol. **131**, 3277–3286
37. Ferrante, G., Ekiel, I. & Sprott, G. D. (1987) Biochim. Biophys. Acta **921**, 281–291
38. Koga, Y., Ohga, M., Nishihara, M. & Morii, H. (1987) Syst. Appl. Microbiol. **9**, 176–182
39. Morii, H., Nishihara, H. & Koga, Y. (1988) Agric. Biol. Chem. **52**, 3149–3156
40. Kramer, J. K. G., Sauer, F. D. & Blackwell, B. A. (1987) Biochem. J. **245**, 139–143
41. Nishihara, M., Morii, H. & Koga, Y. (1989) Biochemistry **28**, 95–102
42. Ferrante, G., Ekiel, I. & Sprott, G. D. (1986) J. Biol. Chem. **261**, 17062–17066
43. Ferrante, G., Richards, J. C. & Sprott, G. D. (1990) Biochem. Cell Biol. **68**, 274–283
44. Ferrante, G., Ekiel, I., Patel, G. B. & Sprott, G. D. (1988) Biochim. Biophys. Acta **963**, 173–182
45. Nishihara, M. & Koga, Y. (1991) Biochim. Biophys. Acta **1082**, 211–217
46. De Rosa, M., Gambacorta, A., Lanzotti, V., Trincone, A., Harris, E. & Grant, W. D. (1986) Biochim. Biophys. Acta **875**, 487–492
47. De Rosa, M., Gambacorta, A. & Nicolaus, B. (1983) J. Membr. Sci. **16**, 287–294

48. De Rosa, M., Esposito, E., Gambacorta, A., Nicolaus, G. & Bu'Lock, J. D. (1980) Phytochemistry **19**, 827–831
49. Langworthy, T. A., Mayberry, W. R. & Smith, P. F. (1974) J. Bacteriol. **119**, 106–116
50. Langworthy, T. A. (1977) J. Bacteriol. **130**, 1326–1332
51. De Rosa, M., Gambacorta, A., Nicolaus, B. & Bu'Lock, J. D. (1980) Phytochemistry **19**, 821–825
52. Thurl, S. & Schafer, W. (1988) Biochim. Biophys. Acta **961**, 233–238
53. Lanzotti, V., De Rosa, M., Trincone, A., Basso, A., Gambacorta, A. & Zillig, W. (1987) Biochim. Biophys. Acta **922**, 95–102
54. Ekiel, I., Sprott, G. D. & Smith, I. C. P. (1986) J. Bacteriol. **166**, 559–564
55. Moldoveanu, N. & Kates, M. (1988) Biochim. Biophys. Acta **960**, 164–182
56. Moldoveanu, N. & Kates, M. (1989) J. Gen. Microbiol. **135**, 2503–2508
57. Poulter, C. D., Aoki, T. & Daniels, L. (1988) J. Am. Chem. Soc. **110**, 2620–2624
58. Nicolaus, B., Trincone, A., Esposito, E., Vaccaro, M. R., Gambacorta, A. & De Rosa, M. (1990) Biochem. J. **226**, 785–791
59. Kushawaha, S. C., Kates, M. & Martin, W. G. (1975) Can. J. Biochem. **53**, 284–292
60. Henderson, R., Jubb, J. S. & Whytock, S. (1978) J. Mol. Biol. **123**, 259–274
61. Teissie, J., Prats, M., Lemassu, A., Stewart, L. C. & Kates, M. (1990) Biochemistry **29**, 59–65
62. Falk, K.-E., Karlsson, K.-A. & Samuelsson, B. E. (1990) Chem. Phys. Lipids **27**, 9–21
63. Stewart, L. C., Yang, P. W., Mantsch, H. H. & Kates, M. (1990) Biochem. Cell Biol. **68**, 266–273
64. Lind, C., Hojeberg, B. & Khorana, H. G. (1981) J. Biol. Chem. **256**, 8298–8305
65. Tsujimoto, K., Yorimitsu, S., Takahashi, T. & Ohashi, M. (1989) J. Chem. Soc. Chem. Commun. 668–670
66. Fredrickson, H. L., de Leeuw, J. W., Tas, A. C., van der Greef, J., LaVos, G. F. & Boon, J. J. (1989) Biomed. Environ. Mass. Spectrom. **18**, 96–105

Biochem. Soc. Symp. **58**, 73–78
Printed in Great Britain

Progress in developing the genetics of the halobacteria

W. Ford Doolittle, Wan L. Lam, Leonard C. Schalkwyk, Robert L. Charlebois, Steven W. Cline and Annalee Cohen

Department of Biochemistry, Dalhousie University, Halifax, Nova Scotia, Canada B3H 4H7

Rationale

Five or six years ago, this laboratory undertook to develop tools which might prove useful for archaebacterial genetics. Progress had been substantial in understanding gene structure and function in archaebacteria, but we were nevertheless limited to the sorts of deductions which could be made from within-kingdom and between-kingdom comparisons of the sequences of genes and regions immediately surrounding them. We were also limited in the genes we could look at to those (i) for which we had protein sequence information, (ii) that by rare good fortune cross-hybridized to eubacterial or eukaryotic homologues, or (iii) that complemented *Escherichia coli* auxotrophic mutations—a technique then and now effective only with AT-rich methanogen DNA. Systems for looking at transcription and translation *in vitro* were primitive. Worse, there were no manipulative genetics, no way to construct strains of desired genetic composition and phenotype, no possibility of mapping mutants and no methods to assess the effects *in vivo* of alterations in gene structure on gene function.

In 1985, Mevarech & Werczberger [1] had reported that by mixing auxotrophs of *Haloferax* (then *Halobacterium*) *volcanii* they could obtain prototrophic 'exconjugants'. They began to look at the underlying biology, while Cline, in our laboratory, had just obtained preliminary evidence that *Halobacterium halobium* could be transfected with DNA from its bacteriophage ϕH. One of these two species seemed like the best candidate for genetic exploitation among the halobacteria, which, as mesophilic aerobes, were clearly the most easily manipulated of the archaebacteria. *Hf. volcanii* grew faster, produced colonies on minimal media, and seemed less seriously infested with transposable elements. When Cline showed that it too could be transfected with ϕH DNA (plaques detectable only on an *Hb. halobium* lawn), the choice became obvious.

We adopted three approaches to *Haloferax* genetics: (i) preparation of a physical map, by top-down and bottom-up methods; (ii) optimization and extension of the DNA uptake system first detected through transfection, and harnessing of it to genetic mapping; and (iii) construction of a shuttle vector and honing of other tools of surrogate genetics and strain construction. Each approach is discussed in turn below. We then present one application of these new methods to basic questions about gene structure and function in archaebacteria.

A physical map [2, 3]

Pulsed-field gel electrophoresis was used to obtain an initial estimate of genome size (about 3.7 million bp) and, towards the end of our mapping endeavour, to close gaps. Most of our map information, however, was obtained through a 'bottom-up' approach, linking cosmid clones together into increasingly longer and fewer 'contigs' on the basis of restriction-site information. Clones were obtained with DNA partially digested with *Mlu*I. *Mlu*I sites occur about once every 4 kbp, so that each insert in the vector Lorist M should produce about ten *Mlu*I fragments, when completely digested. Ten additional ('landmarking') enzymes which, from their observed behaviour with total genomic DNA, should cut each insert on average once, were used in a series of double digests with *Mlu*I. With a genome of this size and the resolution of fragments we obtain, no two clones should produce any *Mlu*I fragments of apparently identical size with the same size double digestion products—that is, no two should bear the same 'landmark'—unless they contain overlapping information.

Contig assembly with randomly chosen clones, aided by searches for new clones by chromosome walking, produced 25 contigs representing about 90 % of the genome. Gaps were closed by using probes corresponding to the ends of contigs in Southern hybridization against blots of pulsed-field gels of digests of total DNA made with infrequently cutting enzymes. All of a minimally overlapping set of 152 cosmids, representing all the cloned DNA, were mapped with six restriction enzymes, and in many cases restriction sites within the uncloned gaps closed by blotting could be localized by a similar procedure. The completed map shows that the *Hf. volcanii* genome is partitioned into five circles: a 2920 kbp circle we call the chromosome and 'plasmids' of 690, 442, 86 and 6.4 kbp, which we call pHV4, pHV3, pHV1 and pHV2.

The six-enzyme restriction map has 903 sites. Restriction-site distribution is not random. In several small, distinct regions of chromosome and plasmid, sites for all six enzymes are many times more frequent than elsewhere. Earlier work by many investigators had shown that halobacterial DNA preparations can nearly always be resolved into two bands: a main band of about 68 mol % G+C (FI) and a satellite of 59 mol % G+C (FII), and that although FII DNA was usually found on plasmids, there might be islands of FII within the chromosome. In *Hb. halobium*, these 'islands' are often the home of insertion sequences, of which this species has a bewildering number—these are of many types and responsible for the high-frequency spontaneous mutation observed in several genes. We showed that several of our regions of frequent restriction sites are in fact FII DNA (by malachite green bisacrylamide column chromatography). We determined, by hybridization to dot blots of the minimal set with probes representing the major insertion sequence family of *Hf. volcanii*, ISH51, that insertion sequences are preferentially, but by no means exclusively, localized in these regions.

Hybridization to dot blots of the minimal set was also used to map a great number of 'real' genes. Among protein-coding genes mapped with specific probes from *Hf. volcanii* or other halobacteria were those for the cell-surface glycoprotein, dihydrofolate reductase, gyrase B, histidinol phosphate aminotransferase, 3-hydroxy-3-methylglutaryl coenzyme A reductase, RNA polymerase subunits B″, B′, A, and C, superoxide dismutase (two genes), all enzymes of tryptophan biosynthesis (in two clusters), and several ribosomal proteins. Six cloned and sequenced tRNA genes, together with genes for 7S RNA and 16S, 23S and 5S ribosomal RNA (two clusters, in that order) were similarly located, and an additional 30–40 loci probably encoding tRNA were found by using bulk end-labelled tRNA as probe. In many cases hybridization was against restriction digests of cosmids from the minimal set, so that resolution in localizing genes was at worst 1 % and at best 0.1 % of the genomic length.

Transformation and genetic mapping [4–6]

Blotting cloned genes to cosmids does not seem much like real genetics, so we undertook to improve and extend the DNA uptake system first demonstrated through transfection, so that we might start to introduce genes into cells. DNA uptake requires spheroplast formation by the simple removal of magnesium or other divalent cations with EDTA, addition of DNA and polyethyleneglycol (of low molecular mass), and spheroplast regeneration on plates.

We routinely obtain 50 % regeneration of spheroplasts, with about 1 % of cells having taken up DNA. Spheroplasts can be stored frozen and cells scraped from plates can be used without re-establishing growth. We have shown transformation of *Hf. volcanii*, *Hb. halobium*, *Haloarcula hispanica* and *Haloarcula vallismortis*, so the method is quite generally applicable. Several species clearly carry restriction systems, but our transformation protocol seems sufficiently efficient that transformants can be easily obtained or quantified. With bacteriophage transfection, this corresponds to about 10^7 plaques/μg of DNA, and similar frequencies can be obtained when cloned DNA is used to complement auxotrophs, so homologous recombination systems must be active. We have so far shown, and used for various purposes: transfection; transformation with native plasmid (scoring by colony hybridization); transformation with chromosomal DNA in the size range 1–100 kbp; transformation with shuttle vectors grown either in *Hf. volcanii* or *E. coli*; transformation with cosmid DNA from *E. coli*; with restriction fragments of cosmid DNA taken from gels; and with double- and single-stranded M13 sequencing templates. Holmes *et al.* [7] had shown that DNA from *dam*$^-$ strains of *E. coli* is not restricted in *Hf. volcanii*, and indeed we obtain 1000-fold better results with *E. coli*-grown DNA when JM110 is used as host.

Armed with transformation, we have begun to accumulate a collection of mutants which we can map by genetic procedures. We have located 140 ethylmethanesulphonate-induced mutants requiring one of 14 amino acids, uracil, adenine, or guanine by transformation with cosmid DNA from the minimal set. We reduced the number of transformations required to about 4000, by first using pools of cosmids (constructed so that each cosmid can be identified by the two pools of which it is a member), and by checking all new mutants with cosmids previously shown to transform other mutants with the same requirement, before resorting to pools.

This has added an additional 35 loci to the map, most represented by several mutants in the collection. All loci are on the 'chromosome', as are all protein-coding genes localized by blotting, except for one of the superoxide dismutase duplicates. Evidence for operon-like organization comes from the observation, for instance, that all of 29 tryptophan auxotrophic alleles map to one of two positions (see below). On the other hand, some biosynthetic pathways controlled by operons in *E. coli* are probably not represented by operons in *Hf. volcanii*—the 23 histidine mutants mapped fall in six unlinked positions.

Transformation with fragments eluted from gels of restriction digests of positive cosmids can localize mutations to within one or two kbp, and followed by sequencing, helps us to find genes. Sequencing of a 1.6 kbp fragment which could transform a histidine auxotroph, for instance, revealed a homologue of the *E. coli hisC* gene, while work with an arginine-auxotroph-transforming fragment has found an operon-like cluster of genes involved in metabolism of that amino acid.

A shuttle vector and other tools [8,9]

To assess the effect of modification *in vitro* of gene structure on function *in vivo*, a vector that can be selected and propagated alternatively in *Hf. volcanii* or *E. coli* was needed. We first showed that pHV2, the endogenous 6.4 kbp plasmid of *Hf. volcanii* DS2, could be cured by growth in ethidium bromide, and re-introduced by transformation (detectable by colony hybridization). Sequencing revealed four likely open reading frames (ORFs) but no clues as to where we might safely disrupt this plasmid. A fortuitously discovered variant of pHV2 with an ISH51 element inserted into one ORF solved this problem, and also provided convenient internal restriction sites for cloning and inactivation of transposition functions.

For a selectable marker, we shotgun cloned into this disrupted plasmid total DNA from spontaneous mutants resistant to the drug mevinolin. This is an inhibitor of HMG-CoA reductase, and of cell growth, for *Hb. halobium* [10], and we found it effective at low concentrations against *Hf. volcanii*, with mutants resistant to much higher levels arising at promisingly low frequency. Subsequent insertion of pieces derived from the *E. coli* vector pBR322, and trimming to remove unnecessary parts of pHV2 and of the mevinolin-resistance determinant gave us pWL102, a vector with half-a-dozen convenient cloning sites, selectable through mevinolin resistance in *Hf. volcanii* and ampicillin resistance in *E. coli*.

Sequencing of the mevinolin-resistance determinant showed it indeed to be the halobacterial homologue of eukaryotic HMG-CoA reductases, and much more like these enzymes in primary sequence than like the one sequenced eubacterial enzyme with this activity, that from *Pseudomonas mevalonii*. Recloning (and sequencing) of the wild-type gene using this as probe revealed that resistance results from a single G $\rightarrow$ T change upstream of the coding region, in what looks like the 'box A' sequence of the canonical archaebacterial promoter sequence (Palm, pp. 79–88 in this book), making a closer approximation to this sequence. Fortuitously, this destroys an *Mlu*I site, and examination, by blotting, of a number of spontaneous mevinolin-resistant mutants shows that all point mutations to resistance destroy this site. About half the mutants are not point mutations, however, but amplifications of the HMG-CoA reductase gene and differing amounts of flanking DNA. We suspect that mutations blocking the ability to bind mevinolin (a competitive inhibitor of HMG-CoA) which retain activity are scarce, and that resistance comes from making more of the wild-type enzyme, either through 'up-promoter' mutations or through gene amplification.

We can use our shuttle vector, or other constructs with the mevinolin-resistance locus (mev^R), for gene disruption experiments and strain construction. When mev^R is inserted in the *trpB* gene, and linear DNA bearing this is transformed into wild-type cells, mevinolin-resistant transformants are often tryptophan auxotrophs. When mev^R flanks a *trp* gene, from which internal sequences are deleted on a circular DNA, this deletion can replace the chromosomal wild-type gene with good efficiency. We are now working on a generalized mutagenesis procedure using such constructs.

A sample application: the tryptophan operon [11]

Genes and proteins of tryptophan biosynthesis have been characterized in many eubacteria and fungi: there is probably more and broader comparative data on this pathway than any other. We have thus undertaken to fully characterize both *trp* gene clusters of *Hf. volcanii*. We have sequenced each—one contains genes homologous to eubacterial *trp C*, *B* and *A* (linked in that order) and the other homologues of *trp D*, *F*, *E* and *G* (in that order). All linked genes overlap by a few base pairs. There are no obvious regions of common sequence upstream of the two clusters, but transcription of the two seems co-ordinately repressed by tryptophan and induced by anthranilate, when these compounds are added to the medium. We have begun a linker-scanning analysis to identify promoter and operator-like sequences.

Quite recently, Meile *et al.* [12] have completed the characterization of the *trp* operon of *Methanobacterium thermoautotrophicum*. Its genes are in the order *EGCFBAD*, quite different from *Hf. volcanii*, and neither archaebacterial operon shares anything with eubacterial operons other than the $B \rightarrow A$ linkage. How selection can keep *trp* genes together, and yet not prevent them from rearranging among themselves, remains mysterious. It will be exciting to determine whether archaebacterial *trp* operons are under the control of repression or attenuation, and whether components of these systems will be homologues of components in eubacterial control systems.

The future

Clearly, we can now approach questions of gene function and regulation in halobacteria with a degree of ease approaching that enjoyed by *E. coli* molecular geneticists for the last decade or two. It might therefore be a good time to pause and decide what questions are important.

References

1. Mevarech, M. & Werzberger, R. (1985) J. Bacteriol. **162**, 461–463
2. Charlebois. R. L., Hofman, J. D., Schalkwyk, L. C. & Doolittle, W. F. (1989) Can. J. Microbiol. **35**, 21–29
3. Charlebois, R. L., Schalkwyk, L. C., Hofman, J. D. & Doolittle, W. F. (1991) J. Mol. Biol. **222**, 509–524
4. Cline, S. W. & Doolittle, W. F. (1969) J. Bacteriol. **169**, 1341–1344
5. Cline, S. W., Schalkwyk, L. C. & Doolittle, W. F. (1989) J. Bacteriol. **171**, 4897–4991
6. Conover, R. K. & Doolittle, W. F. (1990) J. Bacteriol. **172**, 3244–3249
7. Holmes, M. L., Nutall, S. D. & Dyall-Smith, M. L. (1991) J. Bacteriol. **173**, 3807–3813
8. Charlebois, R. L., Lam, W. L., Cline, S. W. & Doolittle, W. F. (1987) Proc. Natl. Acad. Sci. U.S.A. **84**, 8530–8534
9. Lam, W. L. & Doolittle, W. F. (1989) Proc. Natl. Acad. Sci. U.S.A. **86**, 5478–5482
10. Cabrera, J. A., Bolds, J., Shields, P. E., Havel, C. M. & Watson, J. A. (1986) J. Biol. Chem. **261**, 3578–3583
11. Lam, W. L., Cohen, A., Tsouluhas, D. & Doolittle, W. F. (1990) Proc. Natl. Acad. Sci. U.S.A. **87**, 6614–6618
12. Meile, L., Stettler, R., Banholzer, R., Kotok, M. & Leisinger, T. (1991) J. Bacteriol. **173**, 5017–5023

Biochem. Soc. Symp. 58, 79–88
Printed in Great Britain

RNA polymerases and transcription in archaebacteria

W. Zillig, P. Palm, D. Langer, H.-P. Klenk, M. Lanzendörfer, U. Hüdepohl and J. Hain

Max Planck Institut für Biochemie, D-8033 Martinsried, Germany

Introduction

Transcription is a complex process involving proteins such as large transcriptases, composed of several different subunits, and transcription factors as well as DNA templates with signals and regulatory elements. Evolution of such complex systems must coordinately change structural and informational elements. Eukaryal and bacterial transcription machineries differ in a number of details. Eukarya possess three different highly specialized RNA polymerases with a complex subunit composition [1]. For accurate initiation of transcription they require additional factors (reviewed in [2]). In bacteria one type of RNA polymerase is sufficient for transcription of all RNAs. The signal elements for transcription initiation, enhancement and termination differ significantly between eukarya and bacteria. In bacteria the genes are arranged in transcription units allowing coordinate control (reviewed in [3]). As archaea constitute a third distinct domain of organisms [4], a thorough analysis of the archaeal transcription apparatus should help in understanding the phylogenetic relations of the three domains of life.

Structure of archaeal RNA polymerases

The highest resolution of protein structure can be obtained by X-ray crystallography. Crystallization of highly purified RNA polymerase of *Sulfolobus acidocaldarius* has been achieved in our laboratory (E. Hildt, unpublished work). The crystals form either bundles of slightly bent sickles with hexagonal profile or short rods with unsharp surfaces of fracture at their ends. Gel electrophoresis of washed and redissolved crystals shows the typical band pattern of RNA polymerase, proving the authenticity of the crystals. Although the crystals form readily, their size and regularity is still insufficient for good diffraction patterns. Experiments to optimize crystallization conditions and to improve the homogeneity of the polymerase are in progress.

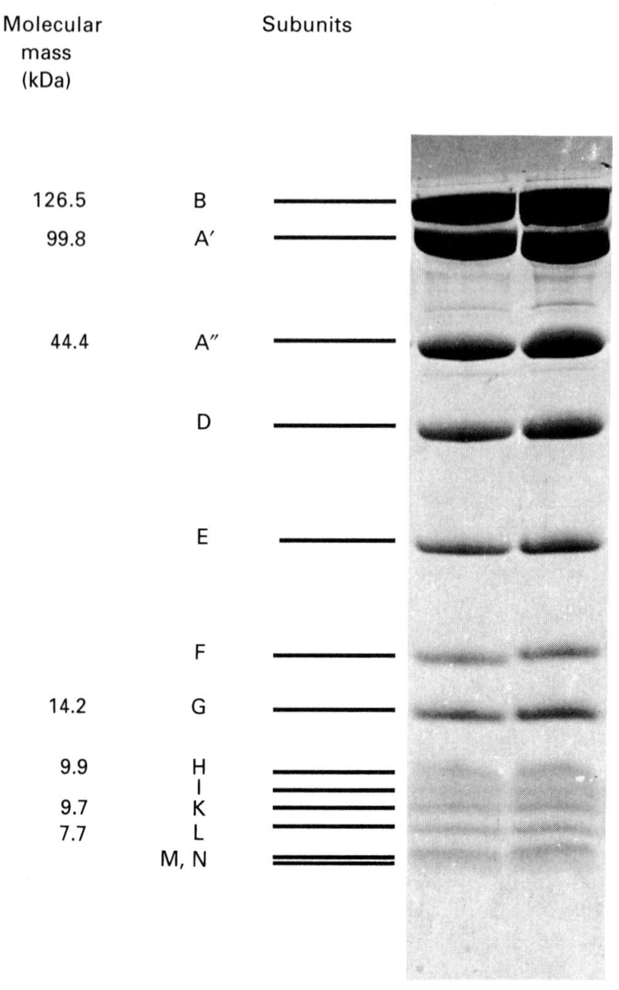

Fig. 1. Subunit composition of RNA polymerase of *Sulfolobus acidocaldarius*. Subunits that have been sequenced are indicated by their molecular masses. Subunits F in the two lanes differ in size by one monomeric unit.

In SDS/PAGE the RNA polymerases of Archaea exhibit a complex subunit pattern [5]. In a gel system specifically designed for high resolution of small peptides [6] the RNA polymerase of *S. acidocaldarius*, a member of the Crenarchaeota, could be separated into 13 subunits named A′, A″, B, D, E, F, G, H, I, K, L, M and N (Fig. 1). From the staining intensities of the bands a one to one ratio of the subunits was deduced. In halophiles and methanogens subunit B is split into two components B′ and B″. The complex subunit pattern of archaea resembles the situation in eukarya whereas the bacteria exhibit a much simpler subunit composition α, β, β', σ [7].

In PAGE subunit F forms a slightly diffuse broadened band, indicating a heterogeneity in mass. In an electrophoresis on a cellulose acetate support [8] a family of six different subspecies of F, with increasing negative charge, can be separated. We assume that a negatively charged oligomer of monomeric units of 200 Da is covalently bound to subunit F. The chemical nature of this oligomer is unclear, but since ^{32}P does not label this subunit, candidates such as phosphates or oligonucleotides are ruled out. In the course of these labelling experiments, however, it has been detected that subunit G is phosphorylated.

Subunits E and I, as well as subunits D and L, form stable complexes that cannot be dissociated by 6 M-urea plus 25% (v/v) formamide and therefore co-migrate in cellulose acetate electrophoresis. Like the polymerase, the DL-complex exhibits a strong yellow colour. The nature of the chromophore is unknown. Heavy metal ions have been excluded.

Enzyme activity can be partially restored after denaturation of RNA polymerase with 6 M-urea and 25% (v/v) formamide by subsequent dialysis against a special reconstitution buffer. Reconstitution of active enzyme from single subunits, that have been separated in cellulose acetate electrophoresis under denaturing conditions, has so far only been achieved with low yield.

Sequences of RNA polymerase subunits

The *N*-terminal amino acid sequences of all *S. acidocaldarius* RNA polymerase subunits (except F and I, which are blocked at the *N*-terminus) were obtained.

The amino acid sequence of the large subunits (A′ and A″, B or B′ and B″) and of subunit H (except of *Methanobacterium autotrophicum*) have been deduced from the corresponding DNA sequences for a number of archaeal RNA polymerases: *Halobacterium halobium* [9], *S. acidocaldarius* [10], *Mb. autotrophicum* [11], *Methanococcus vannielii*, *Thermococcus celer* and *Thermoplasma acidophilum* (I. Arnold-Ammer and P. Palm; H. P. Klenk and V. Schwass, unpublished work). Furthermore, the genes of the *S. acidocaldarius* RNA polymerase subunits G, K and L have been identified by the *N*-terminal amino acid sequence of their encoded proteins. The genes have been cloned and sequenced (D. Langer and F. Lottspeich, unpublished work).

Significant homology exists between archaeal, bacterial and eukaryal large subunits. The archaeal subunits A′ are homologous to the *N*-terminal part of the bacterial subunits β' and the eukaryal subunits A, and the archaeal subunits A″ are homologous to the *C*-terminal part of the bacterial subunits β' and the eukaryal subunits A [9,12,13]. A′ and A″ seem to have arisen from an ancestral uncleaved subunit by the insertion of translational stop and start signals. This event must have occurred in the early archaeal evolution, as all extant archaea exhibit this division as a unique archaeal feature. A formally similar cleavage of the β' subunit of cyanobacteria has been mapped to another position in the middle of the *rpoC* gene and is therefore a distinct evolutionary event [14].

Another cleavage occurred in the evolution of the kingdom of the Euryarchaeota, in which the lowest branches, Thermococcales and the genus *Thermoplasma*, like all Crenarchaeota, possess coherent *B* genes, whereas all methanogens and extreme halophiles contain two fragments B′ and B″, instead. The slightly smaller B″ represents the *N*-terminal part, B′ the *C*-terminal part of the corresponding coherent subunit B. In contrast with the A′/A″ split, this cleavage is a special feature of the higher-branching phyla of the Euryarchaeota.

The archaeal subunit B (B′, B″) exhibits high sequence similarity to all three eukaryal subunits B, and is also related to subunit β of bacteria and chloroplasts.

The archaeal subunit H has no counterpart in the bacterial RNA polymerase, but shows high sequence similarity to the *C*-terminal half of subunit ABC27 [1] of *Saccharomyces cerevisiae* RNA polymerase [15] and subunit HP23 of human RNA polymerase [16]. ABC27 and HP23 are members of a group of eukaryal subunits (ABC27, ABC23 and ABC14.5), which are common to all three types of eukaryal nuclear RNA polymerases (Pol 1, 2 and 3) [1]. No archaeal subunit homologous to the *N*-terminus of ABC27 has been found so far, particularly there is no corresponding open reading frame located upstream of the gene for subunit H.

Similarly high sequence similarities exists between the *S. acidocaldarius* RNA polymerase subunit K and the *C*-terminal half of the *Saccharomyces cerevisiae* RNA polymerase subunit ABC14.5, another member of the group of subunits shared by all three eukaryal RNA polymerases. Subunit K is 83 amino acids long. So far we do not know whether one of the other small subunits of archaeal RNA polymerases is homologous to the *N*-terminus of ABC14.5. We have no explanation for this remarkable tendency of archaea to cut translation units into pieces.

S. acidocaldarius RNA polymerase subunits G and L, the sequences of which have also been determined, are 121 and 67 amino acids long and exhibit no sequence homologies to known eukaryal or bacterial RNA polymerase sequences.

The genes for archaeal RNA polymerase subunits H, B (B″/B′), A′ and A″ form a transcription unit, which corresponds to the *rpoB,C* operon in *Escherichia coli*, except that in *E. coli* the position of the gene for subunit H is occupied by the *rplL* gene, coding for the ribosomal subunit L12 [13]. In *Mb. thermoautotrophicum* the gene for subunit H has not been identified.

Phylogeny

The large subunits of RNA polymerases are particularly well suited for the elucidation of phylogenetic relations between evolutionary distant species. They are fundamental enzymes that are present in all three domains of life and because of their size they provide large data sets. Furthermore patterns of the 20 amino acids of proteins allow a safer, less ambiguous alignment of homologous positions than the pattern of the four nucleotides of nucleic acids.

An alignment of the largest RNA polymerase subunits was established with the aid of the computer programs PROFILE [17] and CLUSTAL [18]. This was corrected for obvious misalignments that arose from the tendency of these algorithms to maximize the total score at the expense of common signatures of several amino acids in length.

A distance matrix was calculated from the pairwise similarities using the formula of Feng *et al.* [19] and a phylogenetic tree was constructed according to Fitch & Margoliash [20] using a computer program designed by J. Felsenstein [21] (Fig. 2). The same branching topology was obtained by several further, independent algorithms of tree construction, including DNA-parsimony, bootstrapped DNA-parsimony and maximum likelihood as accessible in the computer program package

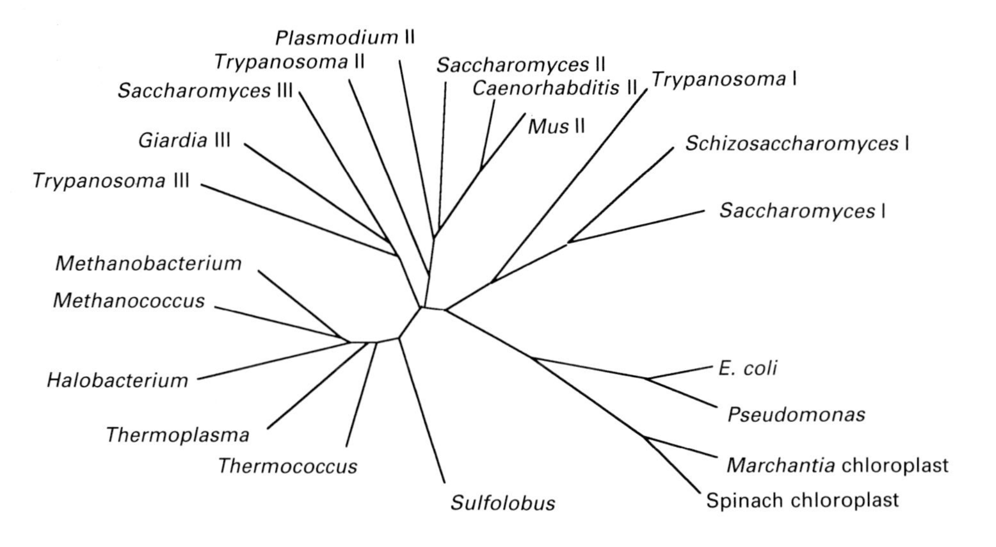

Fig. 2. Phylogenetic tree, calculated from the homologies of the eukaryal subunits A, the bacterial subunits β′ and the archaeal subunits A′ plus A″ of RNA polymerases. A distance matrix was calculated with the formula of Feng *et al*. [19] and the tree constructed according to Fitch & Margoliash [20], using a computer program of Felsenstein [21]. (Sequence references: *Mb. thermoautotrophicum* [11], *H. halobium* [9], *S. acidocaldarius* [10], *E. coli* [38], *Pseudomonas putida* [39], *Marchantia polymorpha* [40], *Spinacia oleracea* [41], *Saccharomyces cerevisiae pol1* [42], *Saccharomyces cerevisiae pol2* and *pol3* [43], *Trypanosoma brucei pol1* [44,45], *Trypanosoma brucei pol2* [46], *Trypanosoma brucei pol3* [47], *Mus musculus* [48], *Schizosaccharomyces pombe* [49], *Plasmodium falsiparum* [50] and *Caenorhabditis elegans* [51]. Sequences of *Giardia lamblia* [52], *Thermococcus celer, Methanococcus vannielii* and *Thermoplasma acidophilum* are unpublished work of this laboratory.)

PHYLIP of J. Felsenstein [21]. It presents the Archaea as a monophyletic–holophyletic group, clearly separated from the eukaryal and bacterial kingdoms, in accordance with the archaebacterial concept of Woese & Fox [4], originally deduced from 16S rRNA sequences. The dendrogram also shows the division of the Archaea into Crenarchaeota [22] (*S. acidocaldarius*) and the Euryarchaeota (*Tc. celer*, *Tp. acidophilum*, *Mc. vannielii*, *Mb. thermoautotrophicus*, and *H. halobium*). The position of *Thermoplasma* is in agreement with its physiological features and the fact that its RNA polymerase subunit B is not split into two subunits.

Another interesting feature of the dendrogram is the fact that the three eukaryal RNA polymerases do not exhibit a common stem, but that Pol1 has a clearly distinct ramification from the branch comprising the Bacteria. From this observation we deduced the hypothesis, that the Eukarya did not originate by a bifurcation from the archaeal branch [23] but rather from fusion events between early Archaea and/or Bacteria. In this concept the ancestor of eukaryal Pol2 and Pol3 was archaeal, whereas Pol1 was imported from the early bacterial lineage [24].

Transcription

Transcription start sites of archaeal genes have been determined for a number of different RNAs. Comparison of the upstream regions of the transcription start sites led to the formulation of a promoter consensus sequence [25–27]. Two sequence motifs have been established, a less conserved box B, which is situated around the transcription start site and a box A centred about 26 nucleotides upstream of the initiation site. The box B consensus motive is (A/T)TG(A/C) containing the transcription start site on the central guanosine or the first adenosine residue. The box A comprises a consensus motif of six nucleotides TTTA(A/T)T. Although bacteria also exhibit two separate motifs in the promoter region, the −35 region and the Pribnow box [3], the archaeal promoter boxes A and B differ fundamentally, both in their sequence and in their location relative to the initiation start. On the other hand there is a high degree of similarity of the archaeal promoter box A to the TATA box of many eukaryal polymerase 2 and 3 promoters, not only in its composition but also in its location 26 nucleotides upstream of the initiation site [28].

A transcription system, *in vitro*, of *Sulfolobus shibatae* was used to evaluate the functional importance of positions in the promoter region [29]. The cloned promoter region of the 16S/23S rRNA gene cluster of the same organism, which conforms perfectly to the archaeal promoter consensus sequence, was used as a template [30]. Transcription efficiency and fidelity of initiation was monitored by S1 nuclease analysis of the transcription products.

Truncation of the template by 5′ deletions of various length up to position −90 did not significantly change the transcription efficiency. Removal of additional sequences in the region between positions −77 to −48 lead to a steady increase of transcription efficiency up to about fourfold, probably owing to the loss of negative regulatory sequences. Further extension of the deletions to positions −35 to −34 drastically reduced the transcription yield and a deletion up to position −27 almost abolished promoter function [31] (Fig. 3*a*).

3′ deletions up to position −2, even including the consensus box B, did not affect transcription efficiency; yet there is strong evidence that a pyrimidine/purine pair represents a start site motif with the initiation occurring at the purine. These experiments define a core promoter region between positions −38 and −2 containing all the information for efficient and specific transcription [31].

A refined study of the core promoter region by the use of clustered and point mutations revealed two sequence elements essential for the 16S/23S rRNA promoter function, a proximal promoter element (PPE) located between positions −12 and −6 and a distal promoter element (DPE) located between positions −36 and −25. The DPE encompasses the box A consensus element (positions −32 to −27). The integrity of this region, conserved between almost all archaea, is essential for transcription efficiency. Most mutations within this element drastically reduced

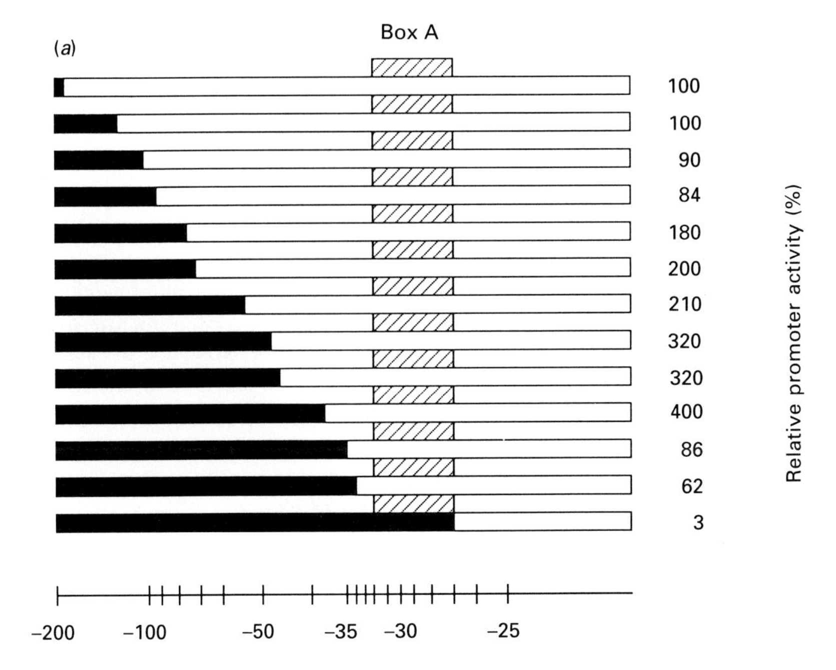

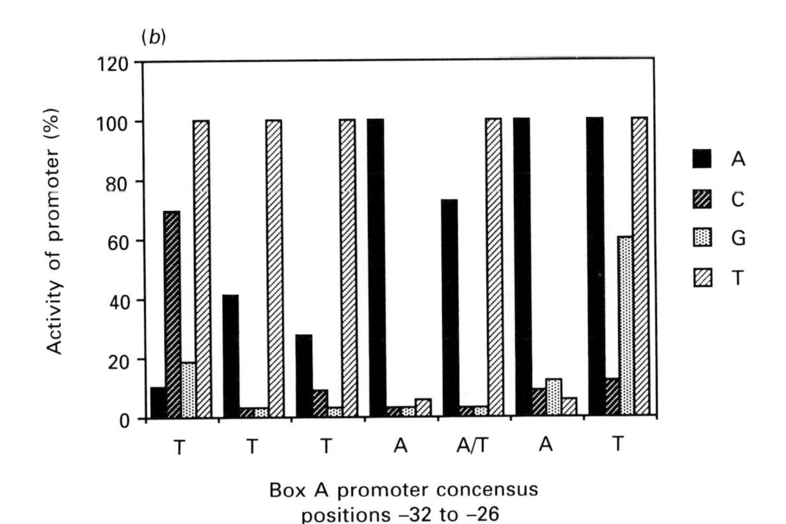

Fig. 3. Activity of the 16S/23S rRNA gene and promoter. (*a*) Template activity, *in vitro*, of the 16S/23S rRNA gene, that has been truncated in its upstream region by deletions of increasing size (represented in log-scale as black bars; numbers give the distance to the initiation site). The relative transcription efficiency was measured by S1 analysis [31]. (*b*) Mutation analysis of the box A of the 16S/23S rRNA promoter. The bars indicate the promoter activity relative to the consensus situation [31].

promoter activity (Fig. 3*b*). The 16S/23S rRNA promoter, which possesses a perfect consensus box A, is the strongest promoter so far analysed, indicating that the consensus sequence is an optimal motif for transcription efficiency.

The PPE has only been defined for the 16S/23S rRNA promoter; a consensus with other promoters could not be established. Mutation of its sequence (AATATGTATA), replacement by its complementary sequence or oligonucleotide sequences composed of either A or T residues, lowered the promoter efficiency drastically.

Variations of the distance between DPE and the initiation site by deletions or insertions led to a shift of the transcription start essentially retaining the wild-type distance. A certain tolerance in the distance was, however, observed in cases where no appropriate pyrimidine/purine pair was present at this position. In the absence of a good box B, start site selection becomes ambiguous though still occurring preferably on a purine following a pyrimidine.

The existence of a universal promoter throughout the archaeal domain has been deduced from the correspondence of the conserved elements. Direct evidence has been drawn from the capacity of *Sulfolobus* RNA polymerase to specifically initiate transcription *in vitro* from two heterologous constitutive promoters operating in *Halobacterium*, that of the plasmid-encoded gene for the major gas vesicle protein [32,33] and that of the predominant early transcript of the *Halobacterium* bacteriophage ϕH [34]. These experiments demonstrate that the basal transcription mechanism is conserved throughout all archaea [35].

Several mapped transcription start regions of highly regulated genes do not exhibit distinct promoter consensus sequences, e.g. the bacterio-opsin gene (*bop*) the bacterio-opsin-related protein gene (*brp*) [36] and the chromosome-encoded gene for the major gas vesicle protein (*c-vac*) [32,33], of *H. halobium*, and the gene of the u.v.-inducible transcript of the *Sulfolobus* virus SSV1 [37]. The *c-vac* gene was also tested as a template in the *Sulfolobus* transcription system *in vitro* and did not stimulate specific RNA synthesis.

The high degree of similarity between archaeal and eukaryal RNA polymerases and the TATA-like box A of archaea emphasizes the close reciprocal relationship between structural and informational elements in the transcription apparatus.

References

1. Mosrin, C. & Thuriaux, P. (1990) Curr. Genet. **17**, 367–373
2. Sentenac, A. (1985) CRC Crit. Rev. Biochem. **18**, 31–91
3. Doi, R. H. & Wang, L.-F. (1986) Nucleic. Acids Res. **14**, 10009–10026
4. Woese, C. R. & Fox, G. F. (1977) Proc. Natl. Acad. Sci. U.S.A. **74**, 5088–5090
5. Zillig, W., Stetter, K. O. & Janekovic, D. (1979) Eur. J. Biochem. **96**, 597–604
6. Schägger, H. & von Jagow, G. (1987) Anal. Biochem. **166**, 368–379
7. Burgess, R. R. (1976) in RNA Polymerase (Losick, R. & Chamberlin, M., eds.), pp. 69–100, Cold Spring Harbor Laboratory, New York
8. Palm, P., Heil, A., Boyd, D., Grampp, B. & Zillig, W. (1975) Eur. J. Biochem. **53**, 283–291
9. Leffers, H., Gropp, F., Lottspeich, F., Zillig, W. & Garrett, R. A. (1989) J. Mol. Biol. **206**, 1–17
10. Pühler, G., Lottspeich, F. & Zillig, W. (1989) Nucleic Acids Res. **17**, 4517–4534

11. Berghöfer, B., Kröckel, L., Körtner, C., Truss, M., Schallenberg, J. & Klein, A. (1988) Nucleic Acids Res. **16**, 8113–8128
12. Pühler, G., Leffers, H., Gropp, F., Palm, P., Klenk, H.-P., Lottspeich, F., Garrett, R. A. & Zillig, W. (1989) Proc. Natl. Acad. Sci. U.S.A. **86**, 4569–4573
13. Zillig, W., Klenk, H.-P., Palm, P., Pühler, G., Gropp, F., Garrett, R. A. & Leffers, H. (1989) Can. J. Microbiol. **35**, 73–80
14. Schneider, G. J. & Haselkorn, R. (1988) J. Bacteriol. **170**, 4136–4140
15. Woychik, N. A., Liao, S.-M., Kolodziej, P. A. & Young, R. A. (1990) Genes Dev. **4**, 313–323
16. Pati, U.-K. & Weissman, S. M. (1989) J. Biol. Chem. **264**, 13114–13121
17. Devereux, J., Haeberli, P. & Smithies, O. (1984) Nucleic Acids Res. **12**, 387–395
18. Higgins, D. G. & Sharp, P. M. (1988) Gene **73**, 237–244
19. Feng, D. F., Johnson, M. S. & Doolittle, R. F. (1985) J. Mol. Evol. **21**, 112–125
20. Fitch, W. M. & Margoliash, E. (1967) Science **155**, 279–284
21. Felsenstein, J. (1984) in Cladistics: Perspectives in the Reconstruction of Evolutionary History (Ducan, T. & Stuessy, T. F., eds.), pp. 169–191, Columbia University Press, New York
22. Woese, C. R., Kandler, O. & Wheelis, M. L. (1990) Proc. Natl. Acad. Sci. U.S.A. **87**, 4576–4579
23. Iwabe, N., Kuma, K., Hasegawa, M., Osawa, S. & Miyata, T. (1989) Proc. Natl. Acad. Sci. U.S.A. **86**, 9355–9359
24. Zillig, W., Klenk, H.-P., Palm, P., Leffers, H., Pühler, G., Gropp, F. & Garrett, R. A. (1989) Endocytobiosis Cell Res. **6**, 1–25
25. Wich, G., Hummel, H., Jarsch, M., Bär, U. & Böck, A. (1986) Nucleic Acids Res. **14**, 2459–2479
26. Reiter, W.-D., Palm, P. & Zillig, W. (1988) Nucleic Acids Res. **16**, 1–19
27. Thomm, M. & Wich, G. (1988) Nucleic Acids Res. **16**, 151–163
28. Bucher, P. & Trifonov, E. N. (1986) Nucleic Acids Res. **14**, 10009–10026
29. Hüdepohl, U., Reiter, W.-D. & Zillig, W. (1990) Proc. Natl. Acad. Sci. U.S.A. **87**, 5851–5855
30. Reiter, W.-D., Palm, P., Voos, W., Kaniecki, J., Grampp, B., Schulz, W. & Zillig, W. (1987) Nucleic Acids Res. **15**, 5581–5595
31. Reiter, W.-D., Hüdephol, U. & Zillig, W. (1990) Proc. Natl. Acad. Sci. U.S.A. **87**, 9509–9513
32. Horne, M. & Pfeifer, F. (1989) Mol. Gen. Genet. **218**, 437–444
33. Englert, C., Horne, M. & Pfeifer, F. (1990) Mol. Gen. Genet. **222**, 225–232
34. Gropp, F., Palm, P. & Zillig, W. (1990) Can. J. Microbiol. **35**, 182–188
35. Hüdepohl, U., Gropp, F., Horne, M. & Zillig, W. (1991) FEBS Lett. **285**, 257–259
36. Betlach, M., Friedman, J., Boyer, H. W. & Pfeifer, F. (1984) Nucleic Acids Res. **12**, 7949–7959
37. Reiter, W.-D., Palm, P., Yeats, S. & Zillig, W. (1987) Mol. Gen. Genet. **209**, 270–275
38. Ovchinnikov, Y. A., Monastyrskaya, G. S., Gubanov, V. V., Gureyev, S. O., Salomatina, I. S., Shuvaeva, T. M., Lipkin, V. M. & Sverdlov, E. D. (1982) Nucleic Acids Res. **10**, 4035–4044
39. Danilkovich, A. V., Borodin, A. M., Allikmets, R. L., Rostapshov, V. M., Chernov, I. P., Azkhikina, T. L., Monastyrskaya, G. S. & Sverdlov, E. D. (1988) Dokl. Biochem. (Engl. Transl.) **303**, 241–245
40. Ohyama, K., Fukuzawa, H., Kohchi, T., Shirai, H., Sano, T., Sano, S., Umesono, K., Shiki, Y., Takeuchi, M., Chang, Z., Aota, S., Inokuchi, H. & Ozeki, H. (1986) Nature (London) **322**, 572–574

41. Hudson, G. S., Holton, T. A., Whitfeld, P. R. & Bottomley, W. (1988) J. Mol. Biol. **200**, 639–654
42. Mémet, S., Gouy, M., Marck, C., Sentenac, A. & Buhler, J.-M. (1988) J. Biol. Chem. **263**, 2823–2839
43. Allison, L. A., Moyle, M., Shales, M. & Ingles, C. J. (1985) Cell (Cambridge, Mass.) **42**, 599–610
44. Jess, W., Hammer, A. & Cornelissen, A. W. C. A. (1989) FEBS Lett. **248**, 123–128
45. Jess, W., Hammer, A. & Cornelissen, A. W. C. A. (1989) FEBS Lett. **258**, 180
46. Evers, R., Hammer, A., Köck, J., Jess, W., Borst, P., Mémet, S. & Cornelissen, A. W. C. A. (1989) Cell (Cambridge, Mass.) **56**, 585–597
47. Evers, R., Hammer, A. & Cornelissen, A. W. C. A. (1989) Nucleic Acids Res. **17**, 3403–3413
48. Ahearn, J. M., Bartolomei, M. S., West, M. L., Cisek, L. J. & Cordon, J. L. (1987) J. Biol. Chem. **262**, 10695–10705
49. Yamagishi, M. & Nomura, M. (1988) Cell (Cambridge, Mass.) **74**, 503–515
50. Li, W. B., Bzik, D. J., Gu, H., Tanaka, M., Fox, B. A. & Inselburg, J. (1989) Nucleic Acids Res. **17**, 9621–9636
51. Bird, D. M. & Riddle, D. L. (1989) Mol. Cell Biol. **9**, 4119–4130

Biochem. Soc. Symp. **58**, 89–98
Printed in Great Britain

Structure, function and evolution of the archaeal* ribosome

Alastair T. Matheson

Department of Biochemistry and Microbiology, University of Victoria, P.O. Box 3055, Victoria, B.C., Canada V8W 3P6

Introduction

Because of its ubiquitous nature the ribosome and its constituent components, the rRNA and r-proteins, have proven to be excellent phylogenetic probes to study molecular evolution. In addition, the comparative data on the structure of specific ribosomal components have identified conserved regions in these molecules which may be involved in the function of this organelle, namely the synthesis of cellular proteins. In this article I will very briefly review the properties of the archaeal ribosome and its rRNA and r-protein constituents and then consider, in more detail, several evolutionary questions, relating to the structure and function of the archaeal r-proteins.

Archaeal ribosomes

The archaeal ribosome (70S) and its subunits (50S and 30S) are similar in size to those of the bacteria and are less complex, in size and in number of constituent molecules, than the eukaryal ribosome (80S). Like the bacterial ribosomes, those from archaea contain three rRNA molecules (23S, 16S, 5S). There are, however, some differences within the archaea as to the number of r-proteins present in its ribosome. Studies by Cammarano *et al.* [2] indicate the presence of two types of

* In this paper I will use the new nomenclature of Woese *et al.* [1] in which Eubacteria, Archaebacteria and Eukaryotes have been raised from kingdom status to a new high level called a domain, and have been renamed Bacteria, Archaea and Eukarya. The Archaea are subdivided into two kingdoms: Euryarchaeota, which contains the extreme halophiles, the methanogens and some extreme thermophiles, and Crenarchaeota, which contains a group of extreme thermophiles whose general phenotype is thought to be similar to the ancestral phenotype of the Archaea.

archaeal ribosomes. One group, which includes the extreme halophiles and most of the methanogens, contains a similar number of r-proteins (approximately 54–56) to that found in bacteria. The second group, which includes the extreme thermophiles and *Methanococcus* spp., contains additional r-proteins for a total of about 28 in the 30S subunit [3] and about 43 in the 50S subunit [4]. In addition, many of these proteins have larger molecular masses than the r-proteins present in the extreme halophiles and methanogens. Since these ribosomes are present primarily in the extreme thermophiles, it is thought that they may reflect the nature of the ancestral ribosomes. By comparison, the eukaryal ribosome contains at least 75–80 r-proteins and these proteins in turn have molecular masses on average that are larger than those found in archaea and bacteria.

Archaeal r-proteins

A large number of archaeal r-proteins have now been sequenced, either directly, using the purified protein and/or indirectly, from the nucleotide sequence of its gene (for reviews, see [5–9]). A recent survey (July, 1991) of the RIBO database (Max-Plank-Institute für Molekulare Genetik, Berlin) indicates that the structure of 64 r-proteins from various extreme halophiles, 27 from *Methanococcus vannielii* and 20 from the extreme thermophile *Sulfolobus*, have been determined and a considerable amount of partial sequence data is also available. When specific r-proteins are compared within the archaea, and when the archaeal r-proteins are compared with the equivalent r-proteins in the bacteria and eukarya, the following conclusions can be made: (1) Specific r-proteins show a great deal of sequence similarity within the archaea and often show features unique to the archaea when compared with those of the equivalent r-protein in eukarya and bacteria. These results support the concept of the archaea as a monophyletic domain (kingdom). (2) The archaeal r-proteins, on the whole, show more sequence similarity to the equivalent proteins in eukarya than to their bacterial counterparts. (3) In general, the size of a specific r-protein is largest in the eukarya, followed by the archaea, which in turn is larger than the equivalent protein in bacteria.

Archaeal r-protein genes

The structure of the r-protein genes in archaea has also provided significant evolutionary data. The structural organization of the genes is similar to that found in bacteria (e.g. *Escherichia coli*) [5–7,9]. The genes are clustered in operons ('*rif*', '*spc*', '*S10*', '*str*') and show an identical order to that found in bacteria. In addition, there are extra opening reading frames (ORFs) dispersed within these operons, or located close to them, which code for r-proteins equivalent to those found in eukarya but showing no sequence similarity to the bacterial r-proteins. The nature of these extra proteins will be discussed later in this article.

The clustering of the r-protein genes into operons, with identical ordering of the genes in both the bacteria and archaea, suggests that the r-protein genes in the primitive ancestral cell were present also in operon-like structures. Herold &

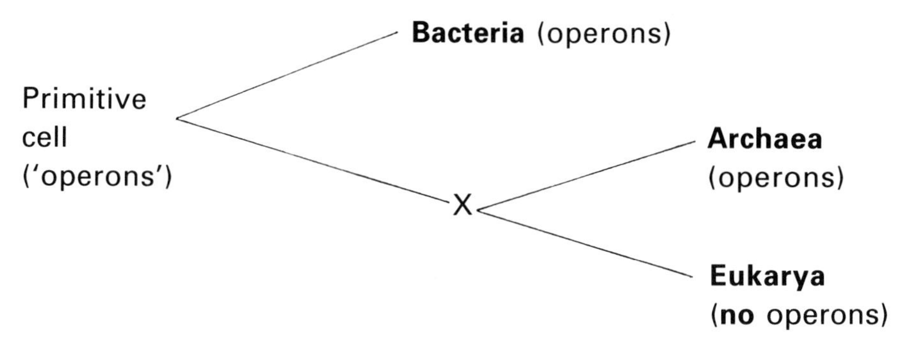

Fig. 1. The evolution of the r-proteins and their genes in bacteria, archaea and eukarya. See text for discussion.

Nierhaus [10] have suggested that the operon structure for the r-protein genes may reflect the assembly steps involved in the formation of the ribosome. As outlined in Fig. 1, when the bacterial line separated from the archaea/eukarya, each group retained the operon structure for their r-protein genes, but the r-proteins within each line could then evolve independently. Later, the archaea and eukarya also separated and the archaea retained the operon gene structure. However, within the evolving nucleated eukaryal cell, in which the assembly of the ribosomes was now separated from the translation process, the clustering of the r-protein genes was no longer required and the genes became scattered throughout the chromosome, often with more than one copy of a specific gene being present. Such a scheme, as outlined in Fig. 1, would explain both the retention of a similar operon structure in the bacteria and archaea, and the observation that the archaeal and eukaryal r-proteins share many structural features not evident in the equivalent bacterial r-proteins.

Archaeal rRNA

A great deal of structural and functional information is now available on the other major component of the ribosome—the rRNA. The primary sequences of a large number of archaeal rRNA molecules have been reported and comparative studies on the rRNA structure from all three domains have resulted in secondary models for the rRNA molecules from both the large and small ribosomal subunits (for review see [11]). Information on the tertiary and quaternary structure of the rRNA is now forthcoming, resulting in models that include the binding sites of the r-proteins in the three-dimensional structure of the rRNA [12–14].

There has been a great deal of discussion on the role of the r-proteins and rRNA in protein synthesis. Previously, it was felt that the r-proteins would prove to contain the catalytic regions for the enzymic steps of translation, and it was thought that the rRNA was present as a structural surface for the proper association of the r-protein molecules. In recent years, however, considerable evidence has accumulated to indicate that the rRNA itself may be involved directly as the catalytic units in translation and, in the extreme case, the r-proteins could be thought to be involved only in the fine tuning of the RNA structure (for review see [15]). A more likely scenario is that both the rRNA and the r-proteins are directly involved in this process. For a detailed review on the structure and function of rRNA the reader is directed to 'The Ribosome' [16], especially to the comments by Peter Moore (p. *xxi*).

Structure and function of the L12–L10 protein complex

One group of proteins that has been thought to play an essential role in translation is the L12–L10 complex [17]. This protein-rich domain makes up the stalk structure present in all large ribosomal subunits. In *E. coli* this domain consists of four copies of an acidic r-protein, L12, present as two dimers which bind by their *N*-terminal domains to r-protein L10, which in turn binds to 23S rRNA. Indirectly involved in this complex is protein L11 whose binding site to 23S rRNA overlaps the binding site for protein L10 [18]. This 23S rRNA region is also the site of binding of thiostrepton which inhibits the elongation factor (EF)-G-dependent hydrolysis of GTP [15]. The protein-rich domain is thought to be involved in factor binding and the 23S rRNA/L11/L10 region may form part of the EF-dependent GTPase centre. Recent studies by Casiano *et al.* [4] have shown a 4:1 complex in *Sulfolobus* equivalent to the $(L12)_4$–L10 in *E. coli* and $(P1)_2(P2)_2$-P0 in eukarya. These results suggest that this quaternary structure has been conserved during evolution and is likely to play an important and direct role in the translocation step.

A great deal of primary sequence data is now available on the L12-equivalent proteins from a wide range of organisms. The bacterial L12 protein is composed of three domains as outlined in Fig. 2. The *N*-terminal region is connected to a *C*-terminal globular region by an Ala/Gly-rich hinge region which gives considerable flexibility to the molecule. The L12 protein is present as two dimers, one of which is thought to form the stalk that is visible in the electron microscope, while the other dimer has been shown to fold towards the body of the 50S subunit [19]. The *C*-terminal globular region has been crystallized and the three-dimensional structure determined [20]. It is this portion of the molecule that is believed to interact with the factors involved in translation.

Considerable sequence data is also available on the equivalent protein in archaea and eukarya. The primary sequences of the equivalent eukaryal and archaeal L12 proteins (L12e) show substantial similarity (see [16] for review) and it is obvious that these molecules arose from a common ancestral gene [21]. Very little sequence similarity is evident when these proteins are compared with the bacterial L12 protein, although an Ala/Gly-rich region similar to the hinge region in the bacterial L12 protein is evident. Several laboratories (see [22] for review) have proposed various models based on the rearrangement of the bacterial L12 protein to obtain maximum sequence similarity between the two groups of L12 proteins.

We had previously postulated [23], from predictive modelling of the *N*-terminal portion of the archaeal/eukaryal L12 protein, that this region might be equivalent to the *C*-terminal globular region in the bacterial L12 protein (see Fig. 2). If this were so, then the archaeal/eukaryal L12 protein would probably bind to the L10 protein via its *C*-terminal region rather than by the *N*-terminal region as in *E. coli*. However, recent experimental results suggest such that a model is incorrect. Antibody studies [24,25] on the eukaryal L12 complex suggest that the *C*-terminal region of these proteins is involved with factor binding, while the *N*-terminal region

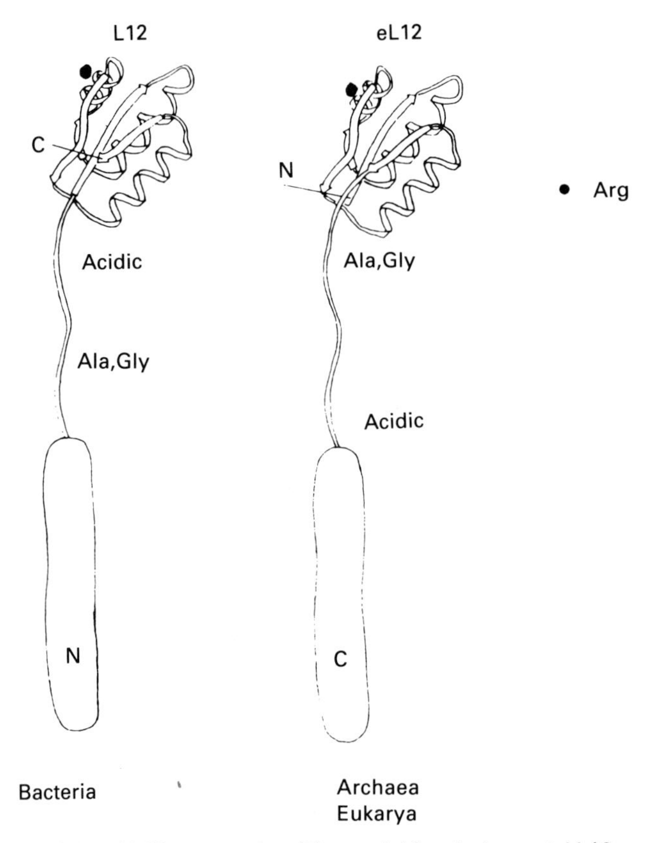

Fig. 2. Models of L12 r-proteins. The model for the bacterial L12 protein is based on a model outlined in [23]. The *N*-terminal domain is linked by a hinge region to the globular *C*-terminal domain whose three-dimensional structure is based on crystallographic data [20]. The proposed model for the archaeal/eukaryal L12 proteins was based on sequence and secondary structure predictions in which the *N*-terminal domain in the archaeal and eukaryal L12 proteins was thought to be equivalent to the *C*-terminal region in the bacterial L12 proteins. Recent experimental data, described in the text, indicates this model is incorrect and the *N*-terminal region in both groups of L12 proteins bind to the L10 protein.

is attached to L10, an arrangement similar to the bacterial L12 complex. In addition, Köpke *et al.* have constructed a series of mutant *Sulfolobus* L12 proteins in archaea that are missing specific regions of the normal protein [26]. These mutant proteins were obtained by site-directed deletions in the *Sulfolobus* L12 gene, followed by overexpression in *E. coli* [27] of the resulting mutant protein. The normal L12 protein was selectively removed from the *Sulfolobus* 50S ribosomal subunit and replaced by the mutant protein. It was found that the *N*-terminal region (residues 1–70) was sufficient for dimerization and incorporation of the L12 protein into the archaeal ribosome. Removal of either residues 1–12 or 1–44 prevents the *N*-terminally truncated mutant protein from binding to the ribosome, indicating that the *N*-

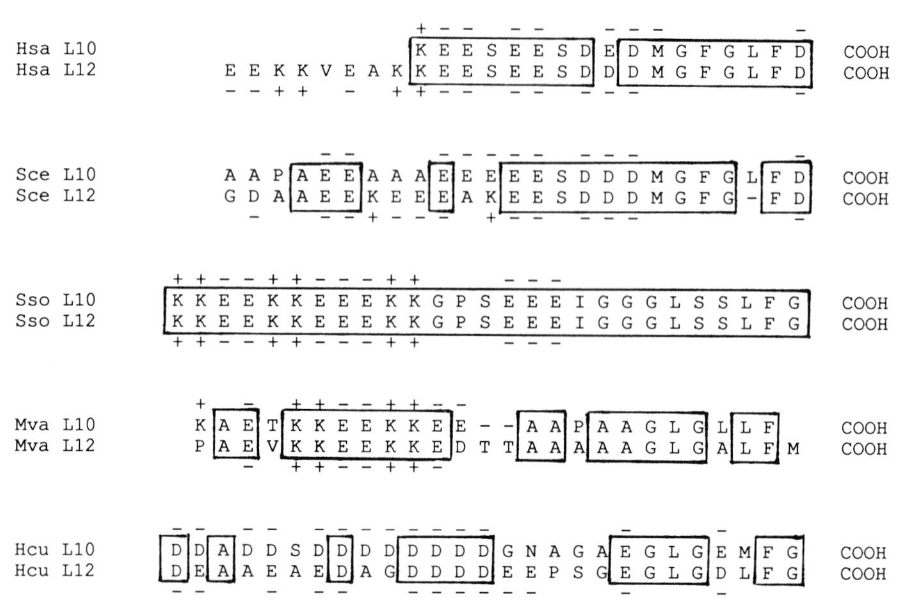

Fig. 3. The C-terminal region in r-proteins L10 and L12 in eukarya [human (Hsa) and yeast (Sce)] and in archaea [*Sulfolobus* (Sso); *Methanococcus* (Mva) and *Halobacterium* (Hcu)]. See [5] for references. Identical residues are boxed and the acidic and basic residues are shown marked − and + respectively. Reproduced from Matheson *et al*. [5], with permission.

terminal region and not the *C*-terminal region was involved in the binding to L10 in archaea. These results argue against models requiring a rearrangement of the L12 protein, in which the *C*-terminal region in the bacterial L12 becomes the *N*-terminal region in the archaeal/eukaryal L12 protein [23].

The above results suggest that the archaeal, eukaryal and bacterial L12 proteins all bind to the L10 protein in the ribosomal core using their *N*-terminal domain even though no significant primary sequence similarity can be detected in this region between the two classes of L12 protein. Experiments using X-ray crystallography or two-dimensional n.m.r. should show whether the two groups of L12 proteins, with different amino acid sequences, will fold to give similar tertiary structures in this region. Such experiments may help to answer the question why the L12 protein, with its essential role in the translation process, has changed so dramatically during evolution. It should also be pointed out that the picture is even more complex in eukarya where there are two types of L12 proteins, P1 and P2 [28], which show a great deal of sequence similarity in the *C*-terminal portion but considerably less in the *N*-terminal portion which binds to r-protein L10.

The $(L12)_4$ L10 domain in eukarya/archaea shows an additional feature which is lacking in the equivalent bacterial complex. The *C*-terminal domains of the eukaryal/archaeal L12 and L10 proteins share a highly conserved, highly charged region as shown in Fig. 3. In *Sulfolobus*, for example, the last 33 residues of both the L12 and L10 proteins are identical [5]. Not only are the amino acid sequences identical but the nucleotide sequences of the two genes are also identical in this region. Other L12–L10 complexes from human, yeast, *Methanococcus* and *Halobacterium* also show substantial conservation of the *C*-terminal regions in these two proteins. The L10 protein in archaea and eukarya is considerably larger than the equivalent protein in bacteria and the small amount of sequence similarity observed between these two classes of L10 protein is mainly evident in the *N*-terminal portion of the larger L10 molecule, suggesting that the highly charged *C*-terminal region found in the archaeal/eukaryal protein may have been added (or lost from the bacterial protein). Assuming the L10 protein binds to 23S rRNA with its *N*-terminal region, as is the case in *E. coli*, this would mean that in the stalk area of the large ribosomal subunit in archaea and in eukarya five copies of the highly charged *C*-terminal region of the $(L12)_4$ L10 complex are available for interaction. Although antibodies specific to the *C*-terminal region inhibit translation in eukarya [25], the role of this highly charged region is still unknown.

Additional r-proteins in archaea

There are a considerable number of ORFs within or close to the operons coding for the archaeal r-proteins. A large number of these ORFs code for r-proteins which in most cases are also present in the eukaryal ribosome. However, the r-proteins coded by these ORFs do not appear to have equivalent proteins in bacteria, or if they do, there is no significant sequence similarity apparent when these r-proteins in archaea are compared with the r-proteins in bacteria. In the case of the thermophiles, additional r-proteins, not present in bacteria, would be expected since as many as 70 r-proteins may be present in *Sulfolobus*. In the halophiles and methanogens, however, the number of r-proteins has been reported [2] to be similar to those found in the bacterial ribosome. A survey of the RIBO database indicates that at least nine r-proteins in the extreme halophiles do not appear to have an equivalent protein in bacteria. If the halophilic and bacterial ribosomes contain approximately the same number of r-proteins, and if the extreme halophile contains proteins not found in bacteria, one would then expect to find r-proteins in bacteria that have no counterpart in archaea. To find out whether this is so we must wait until all the r-proteins in an extreme halophile have been sequenced.

One possible explanation of the situation described above is outlined in Fig. 4. In the ancestral ribosome the rRNA components were probably fragmented and less well structured [29], requiring many polypeptides to hold the catalytic sites in the proper conformation. As the ribosome evolved, the rRNA fragments were combined and streamlined, resulting in a molecule with much more secondary and tertiary structure, so that fewer polypeptides were required to fine tune the catalytic sites. In addition, the polypeptides may have fused into larger proteins containing several 'functional' sites. Since the bacterial line was thought to have separated earlier from the archaeal/eukaryal line (see Fig. 1) the r-protein genes would have evolved independently and the proteins lost during the streamlining of the ribosome may have been different (Fig. 4).

It should be noted that the number of r-proteins assigned to any given ribosome usually is determined by the number of r-proteins separated on a two-

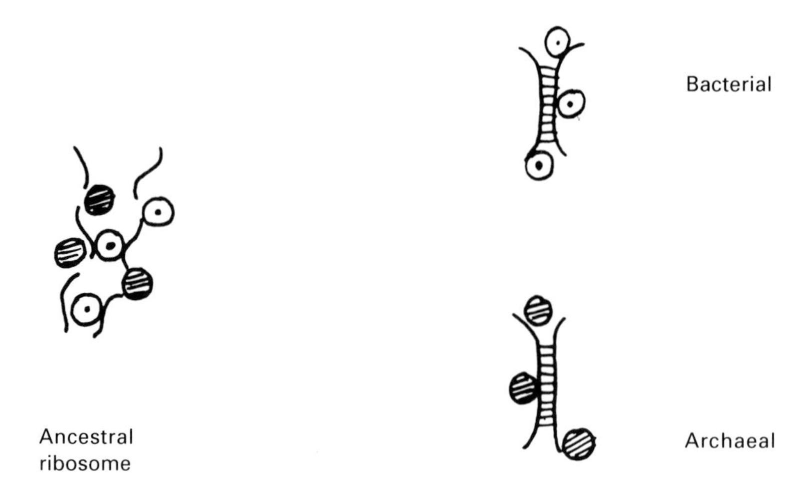

Fig. 4. A proposed model for the streamlining of the rRNA and the resulting loss of r-proteins, during the evolution of the ribosome. See text for discussion.

dimensional polyacrylamide gel. The number, therefore, is very dependent on the ability of the gel to separate the proteins so that care must be taken to evaluate the number of r-proteins present. For example, using the Ström–Visentin two-dimensional system [30] for the separation of the very acidic r-proteins present in the extreme halophiles, it was assumed that these ribosomes contained approximately 55 proteins, which is similar to the number present in *E. coli*. However, Arndt *et al.* [6] have reported additional basic r-proteins in the extreme halophile which would not be detected using the Ström–Visentin system. This would mean that there may be as many as 60 r-proteins in the extreme halophile.

One such basic halophilic protein is L46e [31] which is also present in *Sulfolobus* [32] and is equivalent to a very basic r-protein present in eukarya [33,34]. The *Sulfolobus* r-protein has a pI = 12.9 and shows 46% sequence similarity (identical residues) with the equivalent r-protein in rat liver [34], while the extreme halophile protein has a pI = 11.89 and shows 39% sequence similarity to the eukaryal r-protein. This basic halophilic r-protein is very atypical of halophilic r-proteins in general which are almost all acidic (see [6] for details). The basic nature of this protein in a halophile and the high degree of structural similarity observed between the archaeal and eukaryal L46 proteins, indicates a functional role for this protein in the ribosome. This protein, however, has no equivalent in bacteria, suggesting that it is unique to the archaeal/eukaryal ribosomes.

Several proteins have been reported in archaea which appear to have no counterpart in eukarya and bacteria. One such protein is LX in *Sulfolobus* [35] which is also very basic. Whether this r-protein in unique to the archaea, however, must await the complete sequencing of all the r-proteins in a eukarya.

Within a few years a complete r-protein sequence catalogue should be available for the ribosomes in the three domains and, with the progress being made in crystallizing ribosomal subunits from these domains [36], an accurate, three-dimensional structure for these ribosomes should be possible. The ability to specifically modify r-proteins by site-directed mutagenesis of the r-protein gene, followed by overexpression of the mutant protein in *E. coli* [26,27] and the development of techniques to assemble the archaeal ribosomes *in vitro* [37,38] should provide an opportunity to more fully understand the structure–function relationships of the r-proteins in the archaeal ribosome.

I would like to acknowledge the expert technical assistance of Peter Leggatt, Andrea Louie and Isolde Günther. A special thanks to Celia Ramirez, Andreas Köpke and D. C. Yang for many valuable discussions.

References

1. Woese, C. R., Kandler, O. & Wheelis, M. L. (1990) Proc. Natl. Acad Sci U.S.A. **87**, 4576–4579
2. Cammarano, P., Teichner, A. & Londei, P. (1986) Syst. Appl. Microbiol. **7**, 127–146
3. Schmid, G. & Böck, A. (1982) Mol. Gen. Genet. **185**, 498–501
4. Casiano, C., Matheson, A. T. & Traut, R. R. (1990) J. Biol. Chem. **265**, 18757–18761
5. Matheson, A. T., Auer, J., Ramirez, C. & Böck, A. (1990) in The Ribosome: Structure, Function and Evolution (Hill, W. E., Dahlberg, A., Garrett, R. A., Moore, P. B., Schlessinger, D. & Warner, J. R., eds.), pp. 617–635, Am. Soc. Microbiol., Washington, D.C.
6. Arndt, E., Scholzen, T., Krömer, W., Hatakeyama, T. & Kimura, M. (1991) Biochimie **73**, 657–668
7. Auer, J., Spicker, G. & Böck, A. (1989) J. Mol. Biol. **209**, 21–36
8. Ramirez, C., Shimmin, L. C., Newton, C. H., Matheson, A. T. & Dennis, P. P. (1989) Can. J. Microbiol. **35**, 234–244
9. Wittmann-Liebold, B., Köpke, A. K. E., Arndt, E., Krömer, W., Hatakeyama, T. & Wittmann, H. G. (1990) in The Ribosome: Structure, Function and Evolution (Hill, W. E., Dahlberg, A., Garrett, R. A., Moore, P. B., Schlessinger, D. & Warner, J. R., eds.), pp. 598–616, Am. Soc. Microbiol., Washington, D.C.
10. Herold, M. & Nierhaus, K. H. (1987) J. Biol. Chem. **262**, 8826–8833
11. Gutell, R. R., Weiser, B., Woese, C. R. & Noller, H. F. (1985) Prog. Nucleic Acid Res. Mol. Biol. **32**, 155–216
12. Schuler, D. & Brimacombe, R. (1988) EMBO J. **7**, 1509–1513
13. Stern, S., Weiser, B. & Noller, H. F. (1988) J. Mol. Biol. **204**, 447–481
14. Oakes, M. I., Kahan, L. & Lake, J. A. (1990) J. Mol. Biol. **211**, 907–918
15. Noller, H. R. (1991) Annu. Rev. Biochem. **60**, 191–227
16. Hill, W. E., Dahlberg, A., Garrett, R. A., Moore, P. B., Schlessinger, D. & Warner, J. R. (eds.) The Ribosome: Structure, Function and Evolution, Am. Soc. Microbiol., Washington, D.C.
17. Möller, W. (1974) in Ribosomes (Nomura, M., Tissieres, A. & Lengyel, P., eds.), pp. 711–731, Cold Spring Harbor Laboratory, Cold Spring Harbor, N.Y.
18. Dijk, J., Garrett, R. A. & Müller, R. (1979) Nucleic Acids Res. **6**, 2717–2729
19. Traut, R. R., Tewari, D. S., Sommer, A., Gavino, G. R., Olson, H. M. & Glitz, D. G. (1986) in Structure, Function, and Genetics of Ribosomes (Hardesty, B. & Kramer, G., eds.), pp. 286–308, Springer-Verlag, New York

20. Leijonmarck, M. & Liljas, A. (1988) J. Mol. Biol. **195**, 555–580
21. Matheson, A. T., Louie, K. A. & Henderson, G. N. (1986) Syst. Appl. Microbiol. **7**, 147–150
22. Matheson, A. T. (1986) in The Bacteria, vol. 8. Archaebacteria (Woese, C. R. & Wolfe, R. S., eds.), pp. 345–377, Academic Press Inc., New York
23. Liljas, A., Thirup, S. & Matheson, A. T. (1986) Chem. Scr. **26B**, 109–119
24. Uchiumi, T., Traut, R. R. & Kominami, R. (1990) J. Biol. Chem. **265**, 89–95
25. Vilella, M. D., Remacha, M., Ortiz, B. L., Mendez, E. & Ballesta, J. P. G. (1991) Eur. J. Biochem. **196**, 407–414
26. Köpke, A. K. E., Leggatt, P. A. & Matheson, A. T. (1992) J. Biol. Chem. **267**, 1382–1390
27. Köpke, A. K. E., Hannemann, F. & Boeckh, T. (1991) Biochimie **73**, 647–655
28. Rich, B. E. & Steitz, J. A. (1987) Mol. Cell. Biol. **7**, 4065–4074
29. Gray, M. W. & Schnare, M. N. (1990) in The Ribosome: Structure, Function and Evolution (Hill, W. E., Dahlberg, A., Garrett, R. A., Moore, P. B., Schlessinger, D. & Warner, J. R., eds.), pp. 589–597, Am. Soc. Microbiol., Washington, D.C.
30. Ström, A. R. & Visentin, L. P. (1973) FEBS Lett. **37**, 274–280
31. Bergmann, U. & Arndt, E. (1990) Biochim. Biophys. Acta **1050**, 56–60
32. Ramirez, C., Louie, K. A. & Matheson, A. T. (1989) FEBS Lett. **250**, 416–418
33. Leer, R. J., Van Raamsdonk-Duin, M. M. C., Hagendoorn, M. J. M., Mager, W. H. & Planta, R. J. (1985) Nucleic Acids Res. **13**, 701–709
34. Lin, A., McNally, J. & Wool, I. G. (1984) J. Biol. Chem. **259**, 487–490
35. Ramirez, C., Louie, K. A. & Matheson, A. T. (1991) FEBS Lett. **284**, 39–41
36. Yonath, A., Bennett, W., Weinstein, S. & Wittmann, H. G. (1990) in The Ribosome: Structure, Function and Evolution (Hill, W. E., Dahlberg, A., Garrett, R. A., Moore, P. B., Schlessinger, D. & Warner, J. R., eds.), pp. 134–147, Am. Soc. Microbiol., Washington, D.C.
37. Sanchez, M. E., Urena, D., Amils, R. & Londei, P. (1990) Biochemistry **29**, 9256–9261
38. Altamura, S., Caprini, E. & Londei, P. (1991) J. Biol. Chem. **266**, 6195–6200

Biochem. Soc. Symp. **58**, 99–112
Printed in Great Britain

Chromosome structure and DNA topology in extremely thermophilic archaebacteria

Patrick Forterre, Franck Charbonnier, Evelyne Marguet, Francis Harper* and Gilles Henckes

Institut de Génétique et Microbiologie, C.N.R.S. URA 1354, Université Paris-Sud, 91 405, Orsay Cedex, France and *Institut de Recherches Scientifiques sur le Cancer, B.P. 94, 802 Villejuif Cedex, France

Introduction

How do extreme thermophiles manage to maintain the structural integrity of their chromosome at temperatures near the boiling point of water? This question has fascinated biologists since the discovery of these extremophiles in the early 1970s. The grouping of most extreme thermophiles into the primary kingdom of archaeabacteria added an additional interest to the problem. Phylogenetic data strongly suggest that the common ancestor of all prokaryotes (eubacteria and archaebacteria) was an extreme thermophile [1]. Accordingly, the chromosome of extreme thermophiles today can be considered as reminiscent of the ancestral prokaryotic chromosome. How the adaptation to life at high temperature let its imprinting into the genome of modern prokaryotes is a question of wide interest. We review here our present knowledge of chromosome structure and DNA topology in extremely thermophilic archaebacteria and present recent experimental data from our laboratory.

Chromosome structure and organization

The genomes of two extreme thermophilic archaebacteria, *Thermococcus celer* and *Sulfolobus acidocaldarius*, have been physically mapped recently: they are composed of a single circular chromosome of 1.9 and 3 Mbp, respectively [2,3]. The two genomes of mesophilic archaebacteria analysed up to now (one halophile and one methanogen) are also composed of a single circular chromosome [4,5]. Archaebacterial genomes therefore show striking resemblance to eubacterial genomes.

Histone-like proteins

DNA packaging in eukaryotes occurs via the formation of 'histonic' chromatin, in which the DNA is wrapped around histones to form nucleosomes. In contrast, eubacterial DNA is packaged by free supercoiling and compaction with histone-like proteins. Several putative histones have been isolated from thermophilic archaebacteria (for reviews, see [6,7]). Two of them have been analysed in more detail: the protein HTa, from the moderate thermophile *Thermoplasma acidophilum*, and the protein HMf, from the extreme thermophilic methanogen *Methanothermus fervidus*. The protein HTa condenses the DNA *in vitro* into small particles of about 5 nm (half the diameter of eukaryotic nucleosomes) and protects only 40 bp against nuclease digestion (for review, see [8]). Therefore, it is unlikely that DNA is wrapped around HTa in these particles. HTa exhibits better structural and sequence similarities with the eubacterial protein HU than with eukaryotic histones [9]. In eubacteria, HU probably plays a general role in the modulation of gene expression, via structural alterations of the DNA [10]. In contrast with HTa, protein HMf exhibits much better sequence similarities with eukaryotic histones involved in nucleosome formation than with HU and forms nucleoprotein particles reminiscent of eukaryotic nucleosomes [11]. These particles have been visualized by electron microscopy: if glutaraldehyde fixation is omitted, HMf is removed from the preparation and the particles are replaced by small DNA loops probably corresponding to DNA previously wrapped around HMf. At a low HMf/DNA ratio, HMf binds preferentially to supercoiled DNA, suggesting that DNA is negatively supercoiled in HMf particles. These results suggest that HMf is involved in the formation of nucleosome-like structures in archaebacteria.

The existence of histonic chromatin in archaebacteria is also supported by the observation of Shioda *et al.* [12] who visualized nucleosome-like structures by electron microscopy in the chromatin of *Halobacterium salinarium*. However, these results should be taken with caution since DNA supercoiling can induce the formation of structures resembling nucleosomes (compactosomes) in the non-histonic eubacterial chromatin, according to the method used for samples preparation [13].

Denaturation and thermodegradation of DNA

It is often suggested that histone-like proteins are required in thermophilic archaebacteria to stabilize DNA at high temperatures. Indeed, HTa and HMf, increase the melting temperature (T_m) of linear DNA *in vitro*. However, this is expected for any protein which divides linear DNA into topologically closed domains by preventing rotation of the two DNA strands. Indeed, a topologically closed DNA (in which the number of links between the two DNA strands remains invariable) is intrinsically resistant to denaturation [14]. Since chromosomes are divided by RNA polymerases into multiple topologically closed domains, there is probably no risk of irreversible strand separation in thermophiles, even at very high temperatures. This may explain why there is no correlation between the GC content and the growth temperatures of archaebacteria whereas stable RNAs, which are not topologically closed molecules, are more GC rich in thermophilic than in mesophilic archaebacteria [15,16].

In contrast, DNA at temperatures near 100 °C may be inactivated by

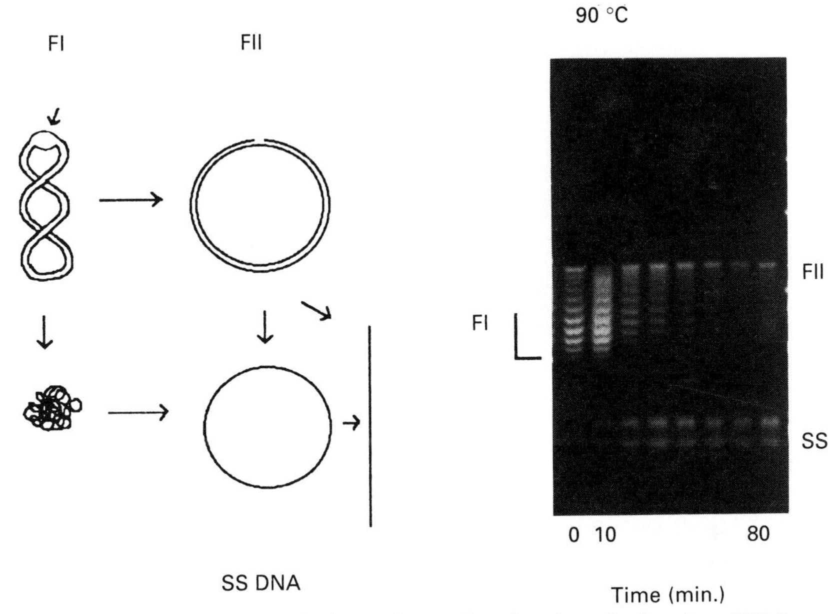

Fig. 1. Thermodegradation of covalently closed circular DNA. Plasmid pTZ18 was incubated for various time periods at 90 °C in the presence of 50 mM-Tris/HCl, pH 7.5 with 6.5% (w/v) ethylene glycol, and run in agarose gel containing ethidium bromide (12 ng/ml) to separate the different topoisomers in form I. The negative superhelicity of these topoisomers increases by increment of one negative superturn from the top to the bottom of the ladder. The two SS bands correspond to linear and circular single-stranded pTZ18.

thermodegradation via depurination, deamination of cytosine and breakage of the 3′–5′ phosphodiester bonds. Thermodegradation of linear DNA was analysed in the years 1960–1970 using time-consuming and poorly informative methods, and without reference to the problem of thermophilic organisms (for reviews, see [7,16]). We recently resumed such investigations on topologically closed DNA, using the simple but powerful technique of agarose gel electrophoresis. Fig. 1 illustrates the harmful effect of DNA cleavage which, in addition to interrupting the continuity of the DNA molecule, removes the topological barriers to denaturation. As a consequence, form I DNA (negatively supercoiled) exposed to high temperature is first transformed into form II by rupture of a single phosphodiester bond, then denatured to give a linear and a circular single-stranded DNA.

Fig. 1 also shows that the most negatively supercoiled topoisomers are the first to be cleaved. This may be explained by the fact that negatively supercoiled DNA contains single-stranded bubbles, and that DNA cleavage occurs more rapidly in single-stranded DNA [17].

Nothing is known about DNA repair and/or protection mechanisms against thermodegradation in extreme thermophiles. Our preliminary results indicate that monovalent and divalent salts partly protect DNA against hydrolysis of the phosphodiester bonds *in vitro* for at least 30 min at 95 °C. This protection is specific

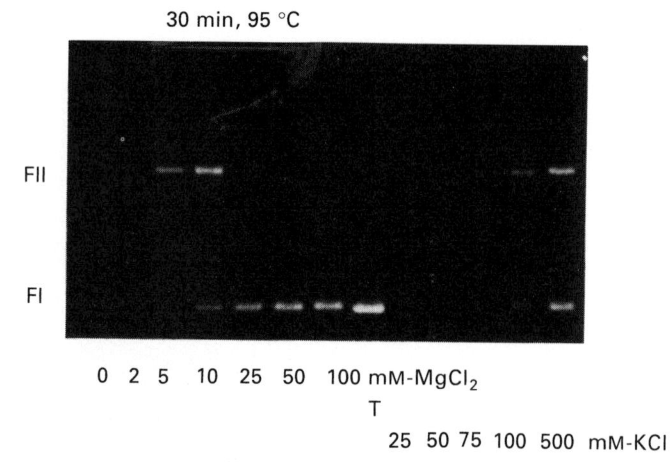

Fig. 2. Protection by salts of DNA against thermodegradation. Plasmid pTZ18 was incubated for 30 min at 95 °C in 50 mM-Tris/HCl, pH 7.5 with various salt concentrations, and then run on agarose gel. Abbreviation used: T, control pTZ18 not incubated at high temperature.

for the salt used; $MgCl_2$ and KCl prevent DNA cleavage at concentrations above 15 mM and 100 mM, respectively (Fig. 2). Zinc and manganese cannot substitute for magnesium (data not shown). The protection of DNA by Mg^{2+} against thermal destruction is in striking contrast with the catalysis of RNA hydrolysis by Mg^{2+} at high temperatures [18].

DNA topoisomerases and DNA topology

Reverse gyrase

Natural DNA isolated from eubacteria and eukaryotes is negatively supercoiled. In eubacteria, negative supercoiling is produced by DNA gyrase, whereas in eukaryotes it is produced by the wrapping of DNA in nucleosomes. Therefore, it came as a surprise when an enzyme producing positive supercoiling was discovered in *S. acidocaldarius* [19,20], for reviews see [7,21]. However, since positively supercoiled DNA exhibits a greater number of topological links between the two DNA strands compared with a negatively supercoiled DNA, it was readily suggested that reverse gyrase activity stabilizes DNA in extreme thermophiles.

Reverse gyrase is a type I DNA topoisomerase. This class of enzyme catalyses the crossing of two DNA strands through each other via a transient single-stranded break, whereas type II DNA topoisomerases catalyse the crossing of two DNA duplexes through each other via a transient double-stranded break. Reverse gyrase resembles eubacterial DNA topoisomerase I by the formation of a reaction intermediate with the protein covalently linked to the 5′ end of the DNA break [22,23], whereas the eukaryotic DNA topoisomerase I binds to the 3′ end.

It has been suggested recently that reverse gyrase could be a combination of a DNA helicase, producing both positive and negative superturns in opening DNA

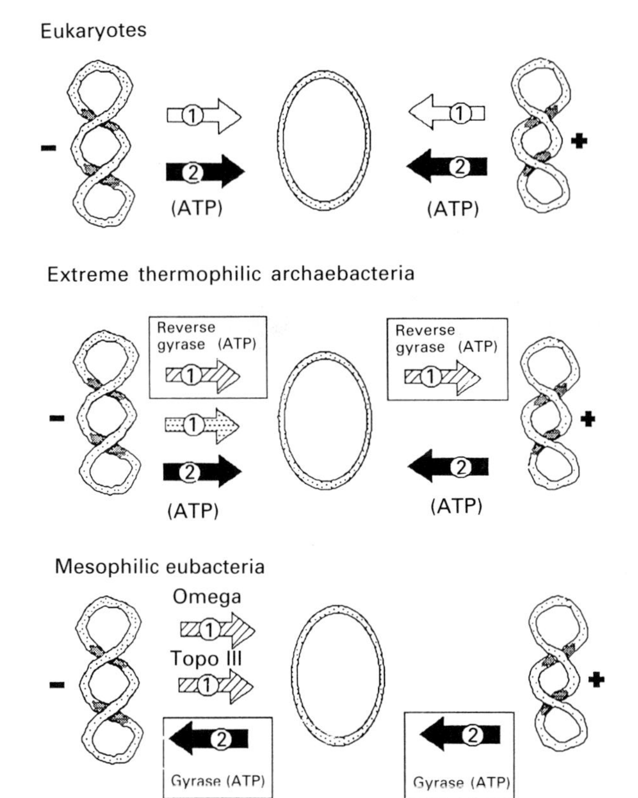

Fig. 3. DNA topoisomerases from the three domains of life. Abbreviations used: Topo III, topoisomerase III; 1, DNA topoisomerase I; 2, DNA topoisomerase II. Arrows with the same shading indicate enzymes which belong to the same evolutionary family, according to their amino acid sequences.

duplexes, and a conventional bacterial DNA topoisomerase I, relaxing only negative superturns [24]. However, purified fractions of reverse gyrase from *S. acidocaldarius* and *Desulfurococcus amylolyticus* contain a single detectable polypeptide of 120–130 kDa [25–27].

Reverse gyrase activity was detected in crude extracts of all extreme and hyperthermophilic archaebacteria tested [28,29], including methanogens [29], and also in extremely thermophilic eubacteria of the order Thermotogales [30]. In contrast, reverse gyrase cannot be detected in extracts of moderately thermophilic and mesophilic archaebacteria [28,29] and eubacteria [28].

A priori, the excellent correlation in all prokaryotes between the presence of reverse gyrase and the extreme thermophilic phenotype strongly supports the hypothesis that reverse gyrase is required to stabilize the DNA double-helical structure at very high temperatures. However, as we previously noticed, any topologically closed DNA (negatively or positively supercoiled) is intrinsically resistant to irreversible denaturation. Since negatively supercoiled DNA contains single-stranded bubbles, which could be the target of DNA cleavage at high temperature (Fig. 1), reverse gyrase could have been required to prevent DNA thermodegradation, by promoting rapid renaturation of melted DNA at high temperature [16]. However, DNA positively supercoiled by reverse gyrase is not more resistant than negatively supercoiled DNA to thermodegradation at physiological salt concentrations (E. Marguet & P. Forterre, unpublished work). The possibility remains that reverse gyrase activity counteracts the formation of transient single-stranded bubbles to prevent abnormal protein/DNA interactions at regulatory sequences and/or to maintain the correct intracellular level of DNA superhelicity at extremely high temperatures.

Other DNA topoisomerases in extremely thermophilic archaebacteria

In addition to reverse gyrase, two DNA topoisomerases have been detected in thermophilic archaebacteria: an ATP-independent activity relaxing only negative superturns and a type II DNA topoisomerase. The ATP-independent topoisomerase is present in crude extracts and partially purified fractions of mesophilic or moderately thermophilic archaebacteria [29,31], and in partially purified fractions of reverse gyrase from extreme thermophiles [20,32]. This enzyme has been separated from reverse gyrase in *D. amylolyticus* [32] and *Sulfolobus shibatae* (M. Nadal & C. Jaxel, personal communication). Slezarev *et al.* showed that this enzyme is a type I DNA topoisomerase and named it 'DNA topoisomerase III' [32]. However, it is not known if this new archaebacterial DNA topoisomerase I is phylogenetically more related to *E. coli* DNA topoisomerase III than to *E. coli* protein ω.

Finally, a type II DNA topoisomerase has been detected in *S. acidocaldarius* [33]. This enzyme has no gyrase activity (unlike the bacterial enzyme) but relaxes either negatively or positively supercoiled DNA, resembling eukaryotic DNA topoisomerase II. However, it may be composed of two subunits, like the bacterial enzyme. Fig. 3 compares the different DNA topoisomerase activities detected up to now in extremely thermophilic archaebacteria with eubacterial and eukaryotic DNA topoisomerases.

Topology of intracellular DNA in extremely thermophilic archaebacteria

SSV1 DNA

The discovery of reverse gyrase raises the question of the sign of DNA supercoiling in extremely thermophilic archaebacteria. Nadal *et al.* [34] first showed that the genome of the u.v.-inducible virus-like particle SSV1 from *Sulfolobus* sp. B12 (now renamed *S. shibatae*, [35]) is positively supercoiled. The DNA of SSV1 isolated from the virus-like particles is highly positively supercoiled, whereas 'intracellular' SSV1 DNA isolated from u.v.-treated cells exhibits a broad distribution of topoisomers, still with a majority of positively supercoiled ones. We recently determined the level of supercoiling of the few SSV1 DNA molecules which can be detected in cells of *S. shibatae* without u.v. irradiation. SSV1 DNA was detected in total DNA extracts by hybridization with ^{32}P-labelled SSV1 probes. Fig. 4 shows that the distribution of SSV1 topoisomers isolated from u.v.-untreated cells is shifted

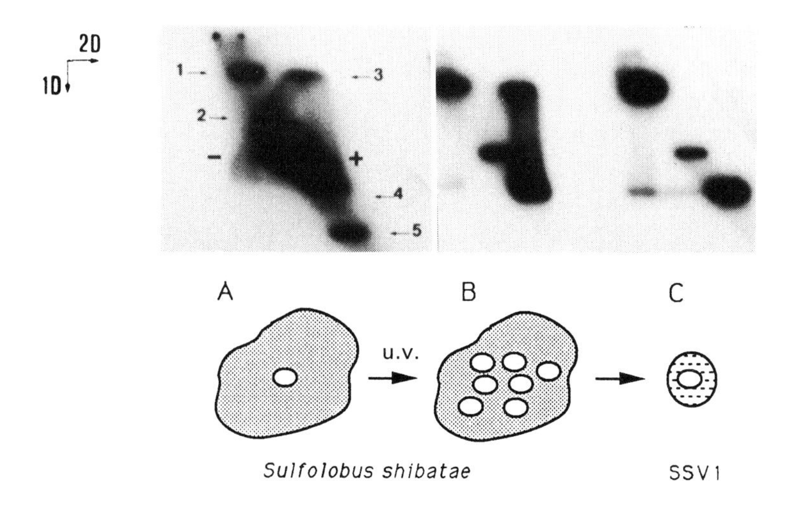

O SSV1 DNA

Fig. 4. Induction of SSV1 positive supercoiling by u.v. irradiation. Panel A: intracellular SSV1 DNA detected in total DNA isolated from *S. shibatae* without u.v.-irradiation. 1, form II; 2, left branch of the arch corresponding to negatively supercoiled DNA (not detected in B and C); 3, top of the arch corresponding to relaxed DNA; 4, positively supercoiled form I; 5, chromosomal integrated form, the big spot in the centre corresponds to linear form III. Panel B: SSV1 DNA isolated from u.v.-irradiated cells, one can see forms I, II and III and the right arch of positively supercoiled DNA. Panel C: SSV1 DNA isolated from viral particles, one can see only forms I, II and III. The method used to detect SSV1 DNA in extracts of u.v.-untreated cells will be described in detail elsewhere; briefly, cells of *S. shibatae* were lysed at room temperature by addition of 1 % (w/v) *N*-lauroyl sarcosine, 1 % (w/v) SDS and treatment with 1 mg/ml proteinase K at 50 °C for 3 h. Total DNA was isolated by ethanol precipitation and RNAase treatment and restricted with *Bg*1*II* and *XbaI* which have several cutting sites on chromosomal DNA but no sites on SSV1 DNA. The products of enzymic digestions were analysed by two-dimensional agarose gel electrophoresis with 5 μg/ml chloroquine in the second dimension. SSV1 DNA in B and C was prepared as in [34]. SSV1 DNA was specifically visualized after Southern blot transfer, using a ^{32}P-labelled SSV1 probe. Abbreviations used: 1D, direction of first electrophoresis run; 2D, direction of second electrophoresis run.

towards the relaxed and negatively supercoiled states (panel A) compared with SSV1 topoisomers isolated from u.v.-treated cells (panel B). This indicates that u.v. induction triggers positive supercoiling of SSV1 DNA.

Plasmid pGT5

Recently, we also performed topological analysis of the first plasmid discovered in an hyperthermophilic archaebacterium. This small multicopy plasmid (3.4 kb), called pGT5 has been isolated from a sulphur-dependent anaerobic archaebacterium (strain GE5) with an optimal growth temperature of 95 °C [36]. This strain was isolated from samples collected at a deep-sea submarine hydrothermal vent. In spite of a strong reverse gyrase activity in strain GE5, the plasmid pGT5 is negatively supercoiled when run on agarose gel at 25 °C. However, its negative superhelical density is lower by a factor of two compared with the superhelicity of eubacterial plasmids. Furthermore, when the effect of temperature on DNA supercoiling is taken into account [37], it appears that pGT5 is relaxed at physiological temperatures (90–102 °C) (F. Charbonnier, G. Erauso, T. Barbeyron, D. Prieur and P. Forterre, unpublished work).

Topology of the chromosome

Recently, we addressed the question of the topological state of the chromosome itself in extremely thermophilic archaebacteria. We first isolated the chromosome of *S. acidocaldarius* with gentle lysis procedures previously used to isolate eubacterial nucleoids. In *E. coli*, two types of nucleoids were obtained according to the lysis temperature or the detergents used [38]: a membrane-bound nucleoid which sediments at about 4000S, and a membrane-free nucleoid of 1300–1600S. Topology of eubacterial chromosomes was determined by analysing changes induced by the DNA intercalating drug ethidium bromide in the sedimentation coefficient of membrane-free nucleoids. Unfortunately, using various lysis conditions, we could only isolate DNA-containing structures with very high sedimentation coefficients, in the range 5500–10 000 S from *S. acidocaldarius*. These 'nucleoïds' contain the S-layer in addition to the chromosome (Fig. 5). Accordingly, they are unsuitable for topological analysis: their sedimentation coefficient does not change, either in the presence of ethidium bromide (as expected for relaxed or negatively supercoiled DNA) or netropsin (as expected for positively supercoiled DNA). Similar results were reported by Collin [39].

Another method used in bacteria and eukaryotes to determine the chromosome superhelicity *in vivo* was to measure the rate of trimethylpsoralen (TMP) photobinding to intracellular DNA [40]. TMP intercalates preferentially into negatively supercoiled DNA, accordingly relaxation of the *E. coli* chromosome by exposure to γ-rays *in vivo* reduces the rate of TMP photobinding. In contrast, there is no change when the same experiment is performed with eukaryotic cells, since negative superturns are constrained around nucleosomes (Fig. 6*a*). We have performed similar experiments with *S. shibatae*. The cells were exposed at room temperature to various doses of γ-rays, then incubated with TMP. TMP molecules were covalently cross-linked to DNA by exposure to u.v., and the DNA was isolated and heat-denatured. The extent of TMP photobinding is deduced from the percentage of TMP cross-linked DNA, i.e. DNA which reanneals rapidly at low temperature. We found that, as in the case of eukaryotes, treatment of *S. shibatae* cells with γ-rays (up to 300 single-stranded DNA breaks) did not reduce the rate of TMP photobinding (Fig. 6*b*). This indicates that the portion of *S. shibatae* DNA accessible to TMP *in vivo* is already relaxed before γ irradiation. However, since TMP binding was performed at 25 °C, this DNA should have been positively supercoiled at the growth temperature of *S. shibatae* (80 °C). This does not exclude that another part of the *S. shibatae* DNA is negatively supercoiled but inaccessible to TMP because it is constrained into nucleosome-like structures.

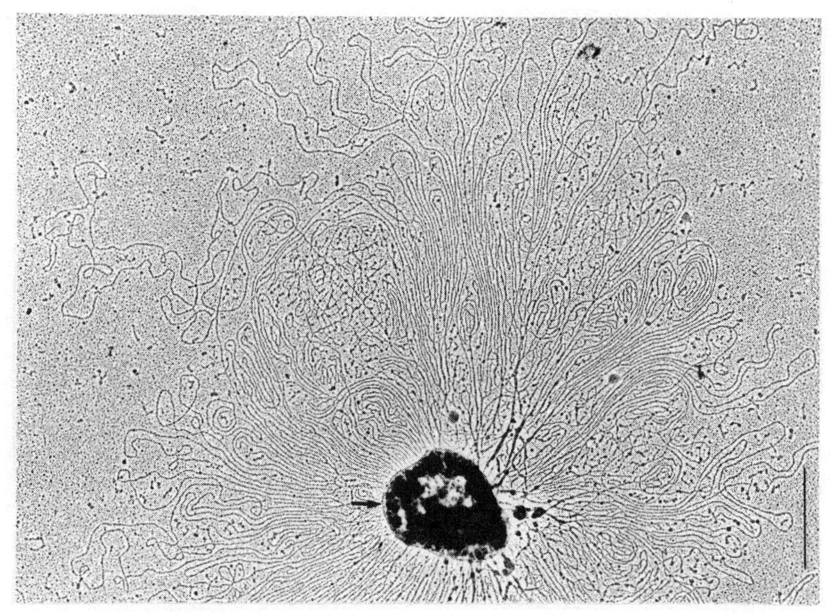

Fig. 5. Electron microscopy of the *Sulfolobus acidocaldarius* 'nucleoid'. The arrow indicates the S-layer, the bar corresponds to 1 μm. Method: cells in exponential growth (strain DSM 639) were quickly cooled to 4 °C and lysed at 60 °C for 10 min in 60 mM-Tris/HCl pH 8.1, 15 mM-EDTA, 1 M-NaCl, 1% (w/v) Brij58 and 0.4% deoxycholic acid. The lysate was sedimented at 4 °C in a sucrose gradient (15/30%) containing 1 M-NaCl. The sedimentation coefficient of the 'nucleoid' was about 7000 S (gradients were calibrated with T4 bacteriophage and DNA content was determined by a fluorescence assay). The 'nucleoid' was spread according to the Kleinschmidt technique.

Fig. 7 shows a simple model for the chromatin from extremely thermophilic archaebacteria: the chromatin (*a*) is divided into regions of free DNA, positively supercoiled by reverse gyrase, and regions of constrained DNA, wrapped in a negative sense around nucleosome-like structures (such as those produced with HMf). When the cells are cooled to room temperature, the free DNA becomes relaxed (*c*). After isolation of chromatin (i.e. removal of nucleosome-like structures) the DNA becomes negatively supercoiled (*d*). DNA superhelicity can be regulated by the activity of DNA topoisomerases (principally reverse gyrase), the amount of available histone-like proteins, and the transcriptional activity which creates both negative and positive supercoiling [41]. U.v. induction of SSV1 (*b*) can promote positive supercoiling by increasing the amount of SSV1 DNA accessible to reverse gyrase in the presence of limiting amounts of histone-like proteins and/or by increasing the number of actively transcribed SSV1 molecules in the absence of sufficient topoisomerase II activity to counteract transcriptional positive supercoiling.

(*a*)

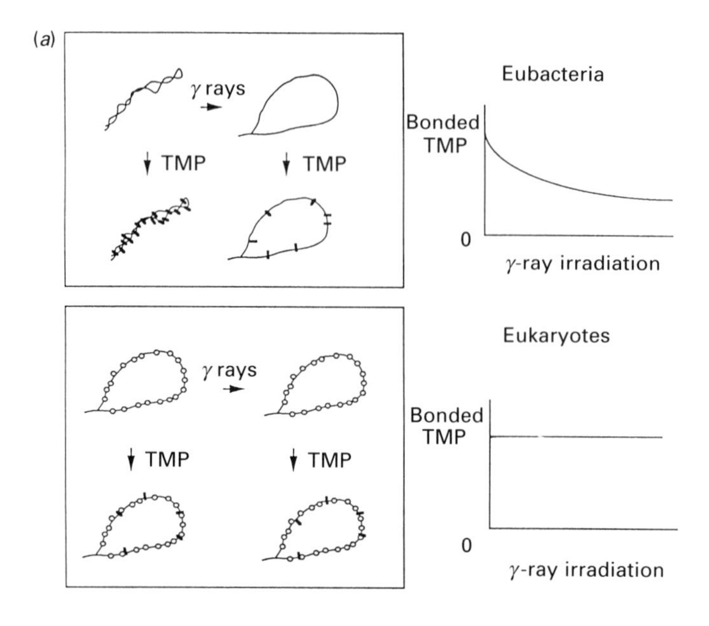

(*b*)

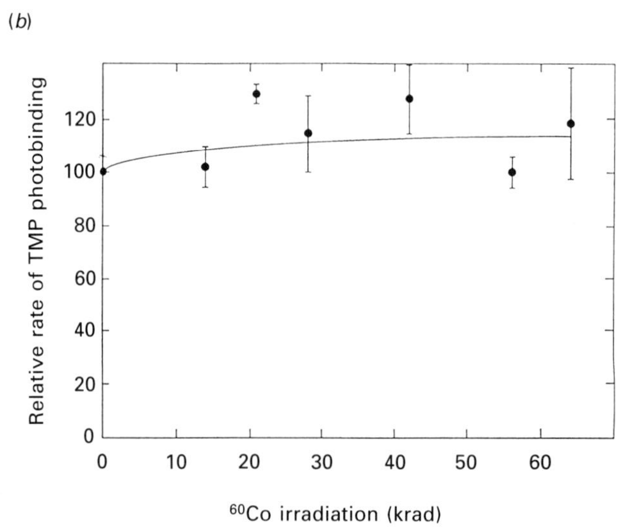

Fig. 6. Determination of chromosome free superhelicity by trimethylpsoralen (TMP) photobinding experiments. (*a*) Principle of the experiment and results obtained in eubacteria and eukaryotes. (*b*) Relative rate of TMP photobinding to DNA in *S. shibatae* cells following γ-irradiation. Cells in exponential growth at 78 °C were quickly cooled and all subsequent steps were performed at room temperature. Cells were exposed to ^{60}Co irradiation in the culture medium (Brock's). The dose rate was 1.4 krad/min and there were about 3.5 single-stranded DNA breaks introduced per krad per genome. Samples removed after various doses of γ-irradiation were treated with TMP and either no u.v.-A or 2 min u.v.-A (at a power density of 2 mW/cm^2). DNA was extracted and purified

Discussion

Comparison of DNA topology in thermophiles and halophiles

In contrast with DNA isolated from extreme thermophilic archaebacteria, plasmids from halophilic archaebacteria are negatively supercoiled, as are those in eubacteria [42]. This suggests that halophilic archaebacteria lack reverse gyrase activity, like other mesophilic archaebacteria. In addition, their DNA topoisomerase II is probably a DNA gyrase, as in eubacteria. Indeed, the genes encoding the two subunits of a DNA topoisomerase II from *Haloferax* strain 2.2 have been cloned recently and sequenced [43] and the amino acid sequences seem to align much better with eubacterial DNA gyrases than with eubacterial and eukaryotic DNA topoisomerase II, which lack gyrase activity. This correlates well with the previous observation that DNA topology in halo-archaebacteria resembles DNA topology in eubacteria rather than that in eukaryotes [44]. As in eubacteria, the DNA gyrase inhibitor novobiocin inhibits DNA replication [42] and induces positive supercoiling of plasmids [45]. These phenomena are not observed in eukaryotes because their type I DNA topoisomerase can relax positive superturns. Interestingly, mechanisms which control DNA could be quite similar therefore in mesophilic archaebacteria and eubacteria on one side (with gyrase activity) and in extremely thermophilic archaebacteria and eubacteria on the other side (with reverse gyrase activity). At the moment, the simplest hypothesis is that archaebacteria and eubacteria share in common a specific thermophilic ancestor with the two activities and that reverse gyrase was lost independently in mesophilic archaebacteria and eubacteria whereas DNA gyrase was lost in extremely thermophilic archaebacteria.

Origin of thermophilic micro-organisms

A current hypothesis is that life arose at high temperature and that the thermophilic ancestor of eubacteria and archaebacteria give rise to all present-day organisms [1]. This was supported by rooting the phylogenetic tree between eubacteria and archaebacteria, based on the analysis of duplicated proteins [46]. However, using protein phylogeny to root the tree of life is not safe; besides the possibility of lateral gene transfer, one cannot be sure that proteins compared in an

---

separately from each sample. DNA preparations were irradiated with 35 kJ of u.v.-A/m^2 to cross-link any TMP mono-adduct in potentially cross-linkable sites, sonicated to reduce the DNA fragments to a single-stranded length of 1000 nucleotides and denatured by heating. For each DNA preparation the fraction of denaturable DNA (without cross-links) was determined using optical methods (hyperchromicity at 260 nm or measurements of the amount of DNA in the single-stranded form and in the double-stranded form after hydroxyapatite chromatography). The mean number of DNA cross-links/fragment (less than 1 DNA cross-link/Kb) was calculated from the fraction of denaturable DNA [50]. For each dose of ^{60}Co irradiation the rate of TMP photobinding was calculated as the number of DNA cross-links formed fragment^{-1} unit^{-1} time of u.v.-A irradiation. Vertical bars indicate the 95% confidence interval of the average of triplicate measurements. All the rates were normalized to the rate obtained without ^{60}Co irradiation, which was taken as 100.

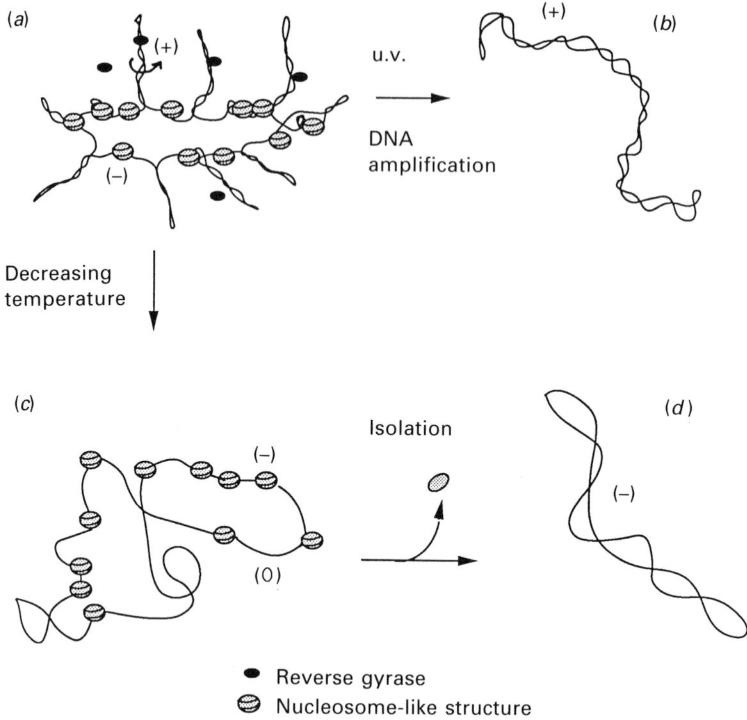

Fig. 7. Model for DNA topology in extremely thermophilic archaebacteria. (*a*) Chromosomal, viral or plasmid chromatin at physiological temperatures. (*b*) SSV1 DNA after u.v. irradiation. (*c*) intracellular DNA after a temperature shift down. (*d*) isolated DNA after deproteinization.

individual tree descend from a single gene in the common ancestor, or from already duplicated genes (see for example [47]). As a consequence, the hypothesis that the thermophilic prokaryotic ancestor emerged from a mesophilic common ancestor to prokaryotes and eukaryotes cannot be rejected. In particular, it could be speculated that the small size of the prokaryotic chromosome and the absence of a nuclear membrane may have originated from reductive adaptation to high temperature [48]. Indeed, the Achilles' heel of organisms living at very high temperature is probably thermodegradation of their macromolecules (as exemplified for DNA in this paper and [16]). Reducing the cell and chromosome sizes permitted a high macromolecular turnover to compensate for thermodegradation. In particular, the absence of a nuclear membrane allowed the rapid use of mRNA, a very important point to counteract the extreme sensitivity of RNA to thermal degradation [49].

Note added in proof

It has been shown recently that DNA could be positively supercoiled *in vitro* around HMf particles [51]. We have also observed that DNA stability is greatly enhanced by phosphate or Hepes buffers.

Work in our laboratory is supported by grants from the Ministère de la Recherche et de la Technologie, the Association de la Recherche contre le Cancer and the Fondation de la Recherche Médicale Française. We thank W. D. Reiter, G. Mirambeau and W. Zillig for help in starting our work with SSV1.

References

1. Woese, C. R. (1987) *Microbiol. Rev.* **51**, 221–271
2. Noll, K. (1989) J. Bacteriol. **171**, 6720–6725
3. Yamagishi, A. & Oshima, T. (1990). Nucleic Acids Res. **18**, 1133–1135
4. Sitzmann, J. & Klein, A. (1991) Mol. Microbiol. **5**, 505–513
5. Charlebois, R., Hofman, J., Schalwyk, L. C., Lam, W. & Doolittle, W. (1989) Can. J. Microbiol. **35**, 21–29
6. Laine, B., Chartier, F., Imbert, B. & Sautière, P. (1990) FEMS Symp. (Belaich, J. P., Bruschi, M. & Garcia, J. L., eds.), pp. 291–301, Plenum Press, New York
7. Forterre, P. & Elie, C. (1992) in The Biochemistry of Archaea (Archaebacteria) (Kates, M., Kushner, D. J. & Matheson, A. T., eds.), in the press
8. Searcy, D. G. (1986) in Bacterial Chromatin (Gualerzi, C. O. & Pon, C. L., eds.), pp. 175–184, Springer Verlag, Berlin & Heidelberg
9. DeLange, R. & Williams, L. (1981) J. Biol. Chem. **256**, 905–911
10 Schmid, M. S. (1990) Cell (Cambridge, Mass.) **63**, 451–453
11 Sandman, K., Krzycki, J. A., Dobrinski, B., Lurz, R. & Reeve, J. N. (1990) Proc. Natl. Acad. Sci. U.S.A. **87**, 5788–5791
12. Shioda, M., Sugimori, K., Shiroya, T. & Takanagi, S. (1989) J. Bacteriol. **171**, 4515–4517
13. Kellenberg, E. (1991) Res. Microbiol. **142**, 229–238
14. Vinograd, J., Lebowitz, J. & Watson, R. (1968) J. Mol. Biol. **33**, 173–197
15. Kaine, B., Schurke, C. & Stetter, K. (1989) Syst. Appl. Microbiol. **12**, 8–14
16. Forterre, P. (1992) in Archaebacteria, Biotechnology Handbooks (Cowan, D., ed.), Plenum Press, in the press
17. Eigner, J., Boedtker, H. & Michaels, G. (1961) Biochim. Biophys. Acta **51**, 165–168
18. Lindahl, T. (1967) J. Biol. Chem. **8**, 1970 1973
19. Kikuchi, A. & Asai, K. (1984) Nature (London) **309**, 677–681
20. Forterre, P., Mirambeau, G., Jaxel, C., Nadal, M. & Duguet, M. (1985) EMBO J. **4**, 2123–2128
21. Kikuchi, A. (1990) in DNA Topology and its Biological Effects (Cozzarelli, N. R. & Wang, J., eds.), pp. 285–298, Cold Spring Harbor Laboratory, Cold Spring Harbor, New York
22. Jaxel, C., Nadal, M., Mirambeau, G., Forterre, P., Takahashi, M. & Duguet, M. (1989) EMBO J. **8**, 3135–3139
23. Kovalsky, O. I., Kozyavkin, S. A. & Slezarev, A. I. (1990) Nucleic Acids Res. **18**, 2801–2805
24. Zhang, H., Hesse, C. B. & Liu, L. F. (1990) Proc. Natl. Acad. Sci. U.S.A. **87**, 9078–9082
25. Nakasu, S. & Kikuchi, A. (1985) EMBO J. **4**, 2705–2710
26. Nadal, M., Jaxel, C., Portemer, C., Forterre, P., Mirambeau, G. & Duguet, M. (1988) Biochemistry **27**, 9102–9108
27. Slezarev, A. I. (1988) Eur. J. Biochem. **173**, 395–399
28. Collin, R. G., Morgan, H. W., Musgrave, D. R. & Daniel, R. M. (1988) FEMS Microbiol. Lett. **55**, 235–239
29. Bouthier de la Tour, C., Portemer, C., Nadal, M., Stetter, K. O., Forterre, P. & Duguet, M. (1990) J. Bacteriol. **172**, 6803–6808

30. Bouthier de la Tour, C., Portemer, C., Huber, R., Forterre, P. & Duguet, M. (1991) J. Bacteriol. **173**, 3921–3923
31. Forterre, P., Elie, C., Sioud, M. & Hamal, A. (1989) Can. J. Microbiol. **35**, 228–233
32. Slezarev, A., Zaitzev, D., Kopylov, V., Stetter, K. & Kozyavkin, S. (1991) J. Biol. Chem. **266**, 12321–12328
33. Kikuchi, A., Shibata, T. & Nakasu, S. (1986) Syst. Appl. Microbiol. **7**, 72–78
34. Nadal, M., Mirambeau, G., Forterre, P., Reiter, W. D. & Duguet, M. (1986) Nature (London) **321**, 256–258
35. Grogan, D., Palm, P. & Zillig, W. (1990) Arch. Microbiol. **154**, 594–599
36. Erauso, G., Charbonnier, F., Barbeyron, T., Forterre, P. & Prieur, D. (1992) Compte Rendu de l'Académie des Sciences, Paris, in the press
37. Depew, R. E. & Wang, J. C. (1975) Proc. Natl. Acad. Sci. U.S.A. **72**, 4275–4279
38. Worcel, A. & Burgi, E. (1972) J. Mol. Biol. **71**, 127–147
39. Collin, R. G. (1991) PhD Thesis, University of Waikato, Waikato.
40. Sinden, R., Carlson, J. O. & Pettijohn, D. (1980) Cell (Cambridge, Mass.) **21**, 773–783
41. Liu, L. & Wang, J. (1987) Proc. Natl. Acad. Sci. U.S.A. **84**, 7024–7027
42. Sioud, M., Possot, O., Elie, C., Sibold, L. & Forterre, P. (1988) J. Bacteriol. **170**, 946–953
43. Holmes, M. & Dyall-Smith, M. (1991) J. Bacteriol. **173**, 642–648
44. Forterre, P., Gadelle, D., Charbonnier, F. & Sioud, M. (1991) in General and Applied Aspects of Halophilic Microorganisms (Rodriguez-Valera, F., ed.), pp. 333–338, Plenum Press, New York & London
45. Sioud, M., Baldacci, G., De Recondo, A. M. & Forterre, P. (1988) Nucleic Acids Res. **16**, 1379–1392
46. Iwabe, N., Kuma, K. I., Hasegawa, M., Osawa, S. & Miyata, T. (1989) Proc. Natl. Acad. Sci. U.S.A. **86**, 9355–9359
47. Benachenou, N. & Baldacci, G. (1992) Mol. Gen. Genet. in the press
48. Forterre, P. (1992) in Frontiers of Life (Thran, T.V., ed.), Editions Frontières, Gif sur Yvette (France), in the press
49. Ginoza, W., Carol, J. H., Vessey, K. B. & Camark, C. (1964) Nature (London) **203**, 606–609
50. Cook, D. N., Amstrong, G. A. & Hearst, J. E. (1989) J. Bacteriol. **171**, 4836–4843
51. Musgrave, D. R., Sandman, K. M. & Reev, J. N. (1991) Proc. Natl. Acad. Sci. U.S.A. **88**, 10397–10401

Biochem. Soc. Symp. **58**, 113–125
Printed in Great Britain

Halophilic malate dehydrogenase—a case history of biophysical investigations: ultracentrifugation, light-, X-ray- and neutron scattering

Henryk Eisenberg

Department of Structural Biology, The Weizmann Institute of Science, Rehovot 76100, Israel

Synopsis

Halophilic malate dehydrogenase (hMDH) from *Haloarcula marismortui* has been isolated, purified and characterized by biochemical and biophysical solution studies. A stabilization mechanism at extremely high concentrations of salt, based on the formation of co-operative hydrate bonds between the protein and hydrated salt ions, was suggested from thermodynamic analysis of native enzyme solutions. Recently the gene coding for hMDH was isolated and sequenced and an active enzyme cloned (F. Cendrin, J. Chroboczek, G. Zaccai, H. Eisenberg and M. Mevarech, unpublished work). A study of the crystal structure of hMDH in a high-salt physiological medium is in progress (O. Butbul-Dym & J. Sussman, personal communication). Here we discuss in depth implications of these recent developments on our earlier results.

Introduction

Halophilic archaebacteria balance the external high salt concentrations of their natural hypersaline medium by accumulating within the cell inorganic ions to concentrations exceeding that of the medium. Therefore, all the cellular components of halophilic archaebacteria have to be adapted to function at these extremely high intracellular salt concentrations. Basic questions within the context of adaptation of the entire organism to extreme concentrations of salt concern the structure, stability and biological activity of proteins and protein–nucleic acid interactions under these unusual conditions [1].

We have, in the past, isolated and purified hMDH from *Haloarcula marismortui* (previously known as *Halobacterium marismortui*) [2], and studied its functional and structural properties by methodologies including ultracentrifugation, light, X-ray and neutron scattering [3–10]. We have concluded that hMDH has a molar mass of 87 000 g/mol, is stable and active at extremely high salt concentrations, and unstable and inactive below 2.5 M-NaCl, for instance [2–5]. In an effort to understand the mechanisms involved in protein stabilization in such extreme environments, we also showed, following thermodynamic analysis by a number of different experimental approaches, that hMDH interacts strongly with about 0.8 g of water/g of protein and about 0.35 g of NaCl/g of protein (or equivalent mole/mole amounts of KCl). The correlation between the solution structure of the enzyme and its stabilization in different environments led to a model in which the main stabilization mechanism consists of the formation of co-operative hydrate bonds between the protein and hydrated salt ions [10]. This interaction was not observed in the case of non-physiological salts and also not at low salt concentrations, following enzyme denaturation [5,8]. We suggested that this hydration mechanism is facilitated, or nucleated, by domains of negative-charge concentration, in a protein carrying excess of negatively charged groups [10]. From the sum of protein, water and salt volumes one can calculate a particle volume, which is in good agreement with volumes calculated from radii of gyration (R_g) and particle shapes independently derived from the angular dependence of X-ray and neutron scattering. The R_g values determined are in considerable excess of the R_g of globular proteins of similar molar mass. Frictional coefficients derived from diffusion coefficients (D) obtained from quasi-elastic light scattering are in agreement with these findings [4].

A consistent pattern thus emerged, and further steps were required to confirm and extend these findings to a molecular basis. To this purpose the sequence of the polypeptide monomer chain was determined, deduced from the nucleotide sequence of the gene coding for the enzyme (F. Cendrin, J. Chroboczek, G. Zaccai, H. Eisenberg and M. Mevarech, unpublished work) and an early preliminary hMDH crystallization and X-ray diffraction study in the non-physiological solvent phosphate [11] was more recently extended to a current study of the crystal structure in a high-salt physiological medium (O. Butbul-Dym & J. Sussman, personal communication).

The amino acid sequence of the enzyme revealed, as expected, an excess of negatively charged acidic groups (glutamate and aspartate over lysine and arginine). Alignment of the amino acid sequence of the enzyme with malate dehydrogenase (MDH) and lactate dehydrogenase (L-LDH) from various sources revealed considerable sequence similarity to the tetrameric LDH from spiny dog-fish (~ 35 %) and to cytoplasmic (~ 21 %) and mitochondrial MDH (~ 28 %). The gene coding for hMDH was cloned into an *Escherichia coli* expression vector and expressed from the strong promoter of the bacteriophage T7. The resulting polypeptide seemed to fold properly, but the enzyme became active only when the salt concentration was increased to high values. Physical properties of the active recombinant hMDH were indistinguishable from the native enzyme. Thus the reduced intensity of neutron scattering extrapolated to zero scattering angle $I_N(0)/c_2$, where c_2 is the protein concentration (g/ml), the R_g values from Guinier plots of the neutron scattering data and the sedimentation coefficient (s) at 4 M-NaCl were identical with those of the native enzyme within experimental error, and crystals identical to those of the native enzyme were obtained (O. Butbul-Dym & J. Sussman, personal communication).

The analysis presented here became necessary owing to the fact that the molar mass M_2 of the monomer polypeptide chain derived from the sequence study was found to be 32816 g/mol, based on a sequence of 304 amino acids, close in composition to the amino acid composition originally described [2]. This value is considerably lower than the physical solution value, if a dimer structure is accepted for the native enzyme ($2 \times 32816 = 65632$, compared with 87000). An additional experimental finding supporting the value derived from the sequence resulted from ion-spray mass spectrometry of the native enzyme. This method, recently extended to the study of the mass of protein subunits [12], claiming high accuracy (yet to be confirmed for this size halophilic entity), yielded a value for the polypeptide chain very close to the value derived from the sequence. A re-examination and comparison of all the physical solution data used in the study of hMDH (*cf.* next section), reveals a stable, slightly revised M_2 (80700 ± 2500 g/mol) which is 23% higher than a sequenced dimer but only 61% of a presumed sequenced tetramer of $M_2 =$ 131264 g/mol. The crystal symmetry and diffraction excludes a trimer, and shows two dimers related by a crystallographic twofold symmetry resulting in a tetramer in the crystal (O. Butbul-Dym & J. Sussman, personal communication). As will be discussed in the following, no concentration-dependent association or dissociation or properties of a mixture were observed in the physical measurements over a broad range of experimental conditions which excludes the existence of association–dissociation equilibria. A tetramer structure which, as indicated above, was not in the range of the solution studies, could provide justification for R_g or D values of a larger protein particle, with smaller water and salt interactions than found previously, yet no justification exists at present for this type of scenario. Cross-linking studies in progress (E. Daniel, personal communication) have so far provided evidence for monomer, dimer and tetramer structures in solution.

The purpose of the present contribution is twofold. When an interesting problem arises in biology, as for instance the adaptation of bacterial life to extreme environments and in the case of our study to high salt concentrations, much effort must be devoted initially to the study and precise characterization of poorly defined proteins in ill-defined circumstances before progress can be achieved in the understanding of the broader problem under study. In the particular case of protein function and structure the final stage of information gathering, often characterized by forgetting the difficult and tortuous path by which the present state of knowledge has been reached, is represented today by modern methods of molecular biology, X-ray diffraction and most recently n.m.r., up to a certain protein size; these often, but not always, make the older approaches obsolete. But there is also the intellectual satisfaction of proving to oneself that the methodology chosen was the correct path by which to reach the present opportunities, which would not have been feasible without the initiative, the hard work, and the ideas which preceded them. Furthermore, to approach new problems, to probe further into the unknown and to create additional new opportunities for the golden tools of modern science, it is of utmost importance in order to avoid future pitfalls, to critically understand the power and the limitations of scientific technologies in a broad sense, as well as their connectivities and complementarities. After all, much of experimental physical and biological science today is black boxes connected to computer terminals, with printed instructions about which buttons to push at what time. Progress is here and we must partake of it, but we must also understand what is behind it, or else rather than climbing new peaks, we might easily find ourselves dropping into deep abysses.

Calculations

A. Average M_2 of hMDH

In this section we calculate an average M_2 of hMDH by sedimentation equilibrium in the ultracentrifuge (SE), sedimentation velocity and diffusion by quasi-elastic laser light scattering (SD) and total-intensity light scattering (LS) measurements, using previously derived experimental quantities only. These calculations are then extended to low-angle X-ray scattering (XS) and small-angle neutron scattering (NS) in the limit of forward (zero-scattering) angle, using calculated values of the partial specific volume $\bar{v}_2$, which has not been experimentally determined in the extreme halophile systems.

Before proceeding to these calculations a thorough check was undertaken to test the correctness of experimental data previously obtained in this system. An essential quantity is the concentration of the pure coenzyme-free enzyme determined from the absorbance at 280 nm. $A_{280}^{0.1\%}$ was independently determined to equal 0.803 [2] (at low salt concentration corrected to high salt concentration) and 0.802 [4] at high salt concentrations by using the same micro Kjeldahl method for nitrogen determination and a value of 17.3 % for the nitrogen content [2]. Considering that these values may have been obtained without sufficient calibration, a revised absorption coefficient $A_{280}^{0.1\%}$ 0.85 ± 0.03 was obtained (F. Bonneté and G. Zaccai, unpublished work). This was done by acid hydrolysis of aliquots of enzyme solutions of known absorbance, including norleucine and phenylalanine added as internal standards, quantitative amino acid analysis, and by averaging amounts of protein obtained from eight 'reliable' amino acid peaks and the amino acid composition as confirmed by the amino acid sequence. Protein concentrations used in previous M_2 determinations will now be divided by 1.06 in view of the increase in absorption coefficient from 0.802 to 0.85.

SE in multicomponent solutions [13,14], at vanishing protein concentrations c_2, is described by

$$\frac{\mathrm{d}\ln c_2}{\mathrm{d}r^2} = \frac{2\omega^2}{\mathrm{RT}} M_2 \left(\frac{\partial \rho}{\partial c_2}\right)_\mu \qquad (1)$$

where M_2 is the molar mass of the protein, r is the distance from the centre of rotation, ω is the angular velocity of the rotor and $((\partial\rho/\partial c_2)_\mu$ is the density increment at constant chemical potential μ of water and salt in solution.

The molar mass M_2 is also derived by SD from the Svedberg equation [13,14],

$$\mathrm{RT}(s/D) = M_2(\partial\rho/\partial c_2)_\mu \qquad (2)$$

where s is the sedimentation coefficient determined in the ultracentrifuge and D the translational diffusion coefficient determined by quasi-elastic light scattering.

From LS measurements at low concentrations, c_2, M_2 is derived by [13,14]

$$N_A \frac{I(0)}{c_2} = \left(\frac{\partial n}{\partial c_2}\right)_\mu^2 M_2 \tag{3}$$

where $I(0)$ is the scattered light intensity extrapolated to zero scattering angle, normalized for wavelength λ and scattering of pure benzene, $(\partial n/\partial c_2)_\mu$ is the refractive index increment at constant chemical potential of diffusible solutes and N_A is Avogadro's number.

From equations (**1–3**) M_2 can be derived by standard experimental procedures without further assumptions. Values of $(\partial \rho/\partial c_2)_\mu$ and $(\partial n/\partial c_2)_\mu$ determined by Mevarech [15] and by Pundak [16] at various NaCl concentrations were increased by multiplying by 1.06 in view of the newly derived absorption coefficient.

The sedimentation coefficient of hMDH in 4 M-NaCl was found to be independent of enzyme concentration between 0.4 and 3 mg/ml (at 260 000 **g** in the analytical ultracentrifuge), decreasing at higher concentrations only [16]. Aggregation was not observed, s in 2.15 M-NaCl at $c_2 = 0.5$ mg/ml is independent of speed of ultracentrifugation between 29 000 and 260 000 **g** [16]. In SE the protein was also found to be homogeneous and experiments at different initial protein concentrations (between 0.1 and 0.8 mg/ml) resulted in identical values of M_2 [15]. From SD in 4.26 M-NaCl buffer and SE in the same buffer M_2 is equal to 78 200 and 79 300 respectively, as determined by Mevarech [15]. From the SD measurements of Pundak [16] in 4 M-NaCl buffer $M_2 = 79\,500$ and from LS in 4 M, 3 M and 2.5 M-NaCl buffer M_2 is equal to 80 100, 85 100 and 83 000 respectively. LS measurements of Leicht [17] in 4.26 M-NaCl buffer were recalculated with $(\partial n/\partial c_2)_\mu$ as determined by Pundak [16], increased by multiplying by 1.06, to give M_2 equal to 78 100 and 82 500 by classical and laser light scattering respectively. The average $M_2 = 80\,700 \pm 2500$. This is still considerably higher than a dimer value from either sequence (65 632) or ion-spray mass spectral analysis (65 402), but considerably lower than tetramer values (131 264 and 130 804 respectively). We have at present no explanation for this discrepancy as the multicomponent solution approach is rigorous and has, in past experiments, led to reliable results, close to sequence data, even at very high salt concentrations. Mevarech & Werber [18] determined in a single experiment a value $13\,600 \pm 600$ from SE at 4.3 M-NaCl for 2Fe-ferredoxin to be compared with 14 727 from the amino acid sequence [19]. They also showed that the M_2 from the empirical SDS/PAGE is about 20 % too high. For the M_2 of *Col* E1 DNA an average of 4.3×10^6 g/mol was obtained from LS and 4.39×10^6 from SD [20]. This corresponds well to the value derived from the sequence 6646 bp of 4.40×10^6 g/mole, for which a definitive value has now been given [21]. More recently [21a] it was shown that M_2 of the monomeric elongation factor Tu from *Ha. marismortui* determined by SD (41 000 g/mole) and NS (48 000 g/mole) is reasonably close to the value obtained from the sequence (45 609 g/mole) [22]; in KCl solutions $B_1 = 0.3$–0.5 and $B_3 = 0.15$–0.25 for this protein, similar but somewhat lower than in hMDH.

In conclusion of this section, properly interpreted experimental results, obtained by a number of investigators over a number of years, with independently purified materials, by a number of experimental techniques, yield an experimental

Table 1. Sedimentation and diffusion coefficients, mass density increments and solute interaction parameters for hMDH at various NaCl concentration (data from [4]).

NaCl conc. (mol/l)	ρ (g/ml)	$s_{20,sol}$ (S)	$D_{20,sol}$ ($\times 10^7$)	$(\partial\rho/\partial c_2)_\mu$ [a]	$-\zeta_1$ [b] $\bar{v}_2 = 0.77$	$\bar{v}_2 = 0.75$	$\bar{v}_2 = 0.73$
5	1.1889	1.44	2.71	0.160	0.399	0.274	0.148
4	1.1560	2.27	3.12	0.220	0.706	0.558	0.409
3.5	1.1379	2.75	3.31	0.251	0.922	0.757	0.688
3	1.1196	3.32	3.54	0.283	1.213	1.026	0.839
2.5	1.1016	4.08	3.77	0.327	1.725	1.508	1.291
2	1.0819	4.72	3.96	0.360	2.357	2.093	1.829
1.5	1.0629	5.51	4.14	0.402	3.505	3.166	2.829
1	1.0432	6.04	4.24	0.430	5.400	4.920	4.434

[a] *cf.* eqn. (**2**); calculated with $M_2 = 80700$.
[b] *cf.* eqn. (**4**); for $\bar{v}_1$ we have assumed unity.

value 80700±2500 g/mole for the M_2 of hMDH, which is 23% higher than the value 65633 derived for a dimer from the amino acid sequence. Here the matter rests unless additional critical experimental evidence can be generated to resolve this unusual discrepancy.

B. Density and refractive index increments and interaction parameters

Having established an average M_2 value in Section A, we can now proceed to calculate $(\partial\rho/\partial c_2)_\mu$ values using eqn. (**2**) and s and D values obtained over a broad range of NaCl concentrations [4]. The density increments can be decomposed into

$$(\partial\rho/\partial c_2)_\mu = (1-\bar{v}_2\rho^0)+\zeta_1(1-\bar{v}_1\rho^0) \tag{4}$$

where $\zeta_1 = (\partial w_1/\partial w_2)_\mu$ is an interaction parameter in multicomponent systems [13,14] in units of grams w_1 of component 1 (H_2O) per grams w_2 of component 2 (protein) and ρ^0 is the solvent density. The interaction term $\zeta_1(1-\bar{v}_1\rho^0)$ can also be expressed in purely symmetric fashion by $\zeta_3(1-\bar{v}_3\rho^0)$ (where component 3 refers to the salt, NaCl), with use of strict thermodynamic arguments [13,14].

To calculate ζ_1 in eqn. (**4**) we must assume a value for the partial volume $\bar{v}_2$, which has not been experimentally determined for this system. In salt-free, or nearly salt-free, solutions, which are required as base states for partial specific volume determination, hMDH is unstable, eliminating the pathway we have used for $\bar{v}_2$ determinations of nucleic acids at high salt concentrations [23]. It may be possible to maintain enzyme stability at low salt concentrations by coenzyme protection, low temperatures, and addition of non-ionic liquid stabilizers, but these difficult procedures have not been explored. The value of $\bar{v}_2$ can be estimated to be 0.73 from the amino acid composition [24] and could be as high as 0.75 or 0.77 if alternate assumptions are made with respect to 'buried' and 'exposed' amino acids [25], or simply as a devils advocate conjecture to estimate variability in calculated ζ_1 values

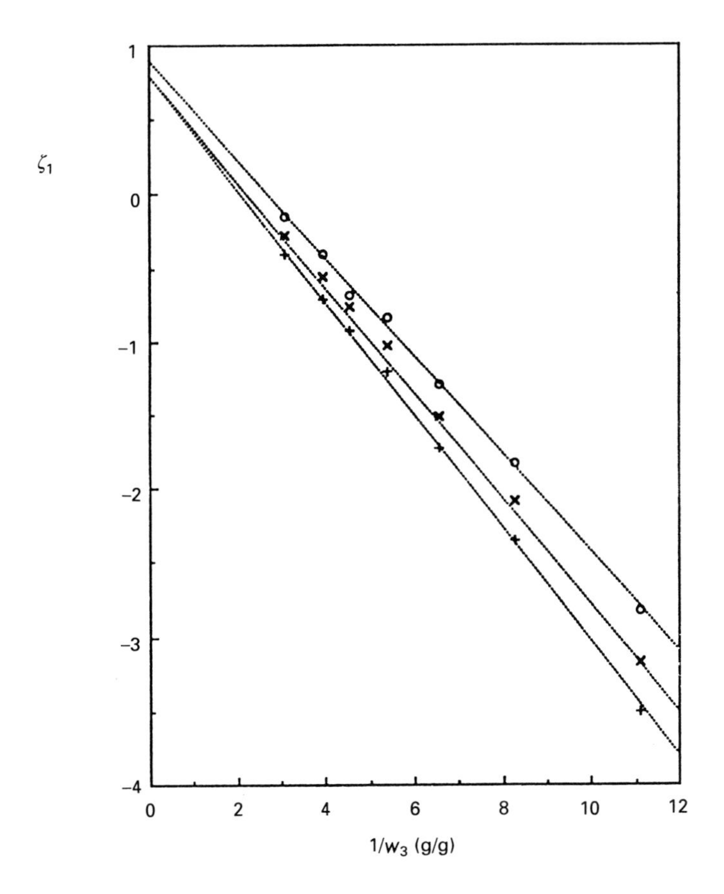

$\bar{v}_2$	0.77 (○)	0.75 (x)	0.73 (+)
B_1	0.8	0.8	0.9
B_3	0.38	0.35	0.33

Fig. 1. Interaction parameters ζ_1 versus $1/w_3$ (see eqn. 5, Table 1).

(Table 1). In Fig. 1 we have plotted, ζ_1 against $1/w_3$ yielding the molecular model 'binding' parameters B_1 and B_3', in g of water and salt respectively per g of protein from the equation [26]

$$\zeta_1 = B_1 - B_3'/w_3 \tag{5}$$

We see that for values $\bar{v}_2 = 0.77$, $B_1 = 0.8$ g/g and $B_3' = 0.38$ g/g, for $\bar{v}_2 = 0.75$, $B_1 = 0.8$ g/g and $B_3' = 0.35$ g/g, whereas for $\bar{v}_2 = 0.73$, $B_1 = 0.9$ g/g and $B_3' = 0.33$ g/g. Thus, neither the thermodynamic parameter ζ_1, nor the molecular model parameters B_1 and B_3' are drastically affected by a significant change in the

Table 2. Experimental and calculated refractive index, density and neutron density increments of hMDH. Abbreviations used: calc., calculated value; exp., experimental value.

NaCl (mol/l)	dn/dc_2[b] (ml/g)	$(\partial n/\partial c_2)_\mu$ (ml/g) calc.[c]	exp.[d]	$d\rho/dc_2$[e]	$(\partial \rho/\partial c_2)_\mu$[f]	$d\rho_N/dc_2$[g] $\times 10^{-9}$ (cm/g)	$(\partial \rho_N/\partial c_2)_\mu$[h]
4	0.160	0.178	0.193	0.133	0.220	16.35	18.33
3.5	0.163	0.184	0.197	0.147	0.251		
3	0.166	0.192	0.204	0.160	0.283	17.02	19.75
2.5	0.169	0.202	0.218	0.174	0.327	17.35	20.70
1						18.35	22.73
0	0.187[a]					19.01	

[a] Calculated [31] from amino acid composition at $\lambda = 589$ nm and adjusted [32] to LS laser $\lambda = 514.5$ nm.
[b] Calculated by eqn. (**6**) from value in water (0.187).
[c] Calculated by eqn. (**7**) from values in column 2, ζ_1 from Table 1 ($\bar{v}_2 = 0.75$), $\bar{v}_1 = 1$ and dn/dc_3 from standard tables.
[d] From [4], corrected to $A_{280}^{0.1\%} = 0.85$
[e] $(1 - \bar{v}_2 \rho)$; $\bar{v}_2 = 0.75$, ρ from Table 1.
[f] From Table 1.
[g] $(b_2 - \bar{v}_2 \rho_N)$, eqn. (**11**); $\bar{v}_2 = 0.75$, b_2 and ρ_N from Table 4.
[h] From Table 4, $\bar{v}_2 = 0.75$.

value of $\bar{v}_2$. Whereas ζ_1 varies strongly with w_3 (or with ρ^0), constant interaction parameters B_1 and B_3' can be obtained from linear plots of $(\partial\rho/\partial c_2)_\mu$ versus ρ^0, if a constant particle model applies [27]. Similar plots are obtained in XS and NS [7].

We emphasize, at this point, the reliability of the M_2 values (cf. Section A) derived from LS [28]. Unlike the density increments $d\rho/dc_2$ in sedimentation, refractive index increments dn/dc_2 are similar for most proteins [29,30] and can also be calculated from the amino acid composition [31]. Corrections can be made for wavelength dispersion [32] and approximate dependence on n [13,14]

$$\left(\frac{dn}{dc_2}\right)_b = \left(\frac{dn}{dc_2}\right)_a - \bar{v}_2(n_b - n_a) \qquad (6)$$

where subscripts a and b refer to values at two different salt concentrations.

The refractive index increment $(\partial n/\partial c_2)_\mu$ is given by [13,14]:

$$\left(\frac{\partial n}{\partial c_2}\right)_\mu = \frac{dn}{dc_2} - \frac{dn}{dc_3}\bar{v}_1 c_3 \zeta_1 \qquad (7)$$

where c_3 is the salt concentration in g/ml. We note that (Table 2*c* and *d*) calculated and experimental values of $(\partial n/\partial c_2)_\mu$ are rather close, and the relative contribution of the interaction term in eqn. (**7**) is considerably less than the corresponding term in the mass increments, eqn. (**4**) (Table 2*e* and *f*). The neutron-scattering density increment terms, eqn. (**11**) (Table 2*g* and *h*) are also close, though, in distinction to $(\partial n/\partial c_2)_\mu$ and $(\partial\rho/\partial c_2)_\mu$, $(\partial\rho_N/\partial c_2)_\mu$ cannot be determined experimentally.

C. Small angle X-ray and neutron scattering

In analogy with forward LS at low concentrations c_2, XS and NS are represented by equations similar to eqn. (**3**)

$$N_A \frac{I_{el}(0)}{c_2} = \left(\frac{\partial \rho_{el}}{\partial c_2}\right)_\mu^2 M_2 \tag{8}$$

and

$$N_A \frac{I_N(0)}{c_2} = \left(\frac{\partial \rho_N}{\partial c_2}\right)_\mu^2 M_2 \tag{9}$$

where $I_{el}(0)$ is the forward scattering of X-rays, calibrated by the scattering of a protein of known mass and known electron density increment $(\partial \rho_{el}/\partial c_2)_\mu$ [33], and $I_N(0)$ is the forward scattering of neutrons, calibrated by the incoherent scattering of water under the same experimental conditions [34].

Neither

$$\left(\frac{\partial \rho_{el}}{\partial c_2}\right)_\mu = (l_2 - \bar{v}_2 \rho_{el}) + \zeta_1 (l_1 - \bar{v}_1 \rho_{el}) \tag{10}$$

nor

$$\left(\frac{\partial \rho_N}{\partial c_2}\right)_\mu = (b_2 - \bar{v}_2 \rho_N) + \zeta_1 (b_1 - \bar{v}_1 \rho_N) \tag{11}$$

(where the l_i and b_i are electron and neutron scattering parameters calculated from the chemical composition of the components and ρ_{el} and ρ_N are the electron and neutron scattering densities of the solvents) are experimentally measurable quantities. They can be evaluated by combining eqns. (**4**) and (**10**) and (**11**), to yield [33],

$$\left(\frac{\partial \rho_{el}}{\partial c_2}\right)_\mu = l_2 + \left(l_i - \frac{\rho_{el}}{\rho}\right)\zeta_1 - \frac{\rho_{el}}{\rho}\left(1 - \left(\frac{\partial \rho}{\partial c_2}\right)_\mu\right) \tag{12}$$

for X-rays, and

$$\left(\frac{\partial \rho_N}{\partial c_2}\right)_\mu = b_2 + \left(b_1 - \frac{\rho_N}{\rho}\right)\zeta_1 - \frac{\rho_N}{\rho}\left(1 - \left(\frac{\partial \rho}{\partial c_2}\right)_\mu\right) \tag{13}$$

for neutrons. Tables 3 and 4 indicate that, whereas in the case of X-rays $(\partial \rho_{el}/\partial c_2)_\mu$ increases with decreasing values of $\bar{v}_2$ assumed for the calculation of ζ_1, in the case of neutrons the opposite is true. This is a direct result of the values the parameters l_1, l_2 and ρ_{el} on the one hand, and b_1, b_2 and ρ_N assume in these equations.

Conclusions

We are now reaching the conclusion of this analysis on the structure of hMDH in extreme saline environments. It has been claimed in the past that, given three points (or parameters) one can build an elephant. We hope to have shown here that this is precisely the opposite in our approach. We have used a number of powerful

Table 3. X-ray forward scattering, electron density increments and molar mass of hMDH at various NaCl concentrations (data from [6]). Abbreviation used: Avg, average value.

NaCl (mol/l)	ρ_{el} ($\times 10^{-23}$ e/cm^3)	$I_{el}(0)/c_2$ ($\times 10^{-23}$e^2/g)	$(\partial\rho_{el}/\partial c_2)_\mu$ [a] ($\times 10^{-23}$e/g)	M_2 [d] (g/mol)	$(\partial\rho_{el}/\partial c_2)_\mu$ [b] ($\times 10^{-23}$e/g)	M_2 [d] (g/mol)	$(\partial\rho_{el}/\partial c_2)_\mu$ [c] ($\times 10^{-23}$ e/g)	M_2 [d] (g/mol)
4.0	3.75	6093	0.630	92500	0.644	88500	0.659	84500
3.0	3.63	7654	0.783	75200	0.802	71700	0.821	68400
2.0	3.55	13680	0.985	85000	1.001	82200	1.017	79700
1.5	3.50	12850	1.085	65700	1.102	63700	1.119	61800
1.0	3.45	16090	1.151	73100	1.169	70900	1.186	68900
			Avg	78300 $\pm$ 10500		75400 $\pm$ 990		72700 $\pm$ 9200

[a] *cf.* eqn. (10); ζ_1 calculated with $\bar{v}_2 = 0.77$ (Table 1); $I_1 = 3.343 \times 10^{23}$e/g; $I_2 = 3.23 \times 10^{23}$ e/g
[b] *cf.* eqn. (10); ζ_1 calculated with $\bar{v}_2 = 0.75$ (Table 1).
[c] *cf.* eqn. (10); ζ_1 calculated with $\bar{v}_2 = 0.73$ (Table 1).
[d] *cf.* eqn. (8).

Table 4. Neutron forward scattering, neutron scattering density increments and molar mass of hMDH at various NaCl concentrations (data from [7]). Abbreviation used: Avg, average

NaCl (mol/l)	ρ_N ($\times 10^{-9}$/cm^2)	$I_N(0)/c_2$ (cm^2/g)	$(\partial\rho_N/\partial c_2)_\mu$ [a] ($\times 10^{-9}$ cm/g)	M_2 [d] (g/mol)	$(\partial\rho_N/\partial c_2)_\mu$ [b] ($\times 10^{-9}$ cm/g)	M_2 [d] (g/mol)	$(\partial\rho_N/\partial c_2)_\mu$ [c] ($\times 10^{-9}$ cm/g)	M_2 [d] (g/mol)
4	−2.07	54.67	18.90	92200	18.33	98000	17.76	104400
3	−2.96	58.81	20.31	85900	19.75	90800	19.19	96100
2.5	−3.40	65.62	21.13	88500	20.70	92200	20.15	97400
1	−4.73	82.40	23.25	91800	22.73	96100	22.20	100700
			Avg	89600 ± 3000		94300 ± 3400		99700 ± 3700

[a] *cf.* eqn. (11); ζ_1 calculated with $\bar{v}_2 = 0.77$ (Table 1); $b_1 = -5.62 \times 10^9$ cm/g, $b_2 = 14.8 \times 10^9$ cm/g.
[b] *cf.* eqn. (11); ζ_1 calculated with $\bar{v}_2 = 0.75$ (Table 1).
[c] *cf.* eqn. (11); ζ_1 calculated with $\bar{v}_2 = 0.73$ (Table 1).
[d] *cf.* eqn. (9).

experimental methods which we have juxtaposed without making any arbitrary assumptions; and in one case, in which a parameter, $\bar{v}_2$, was adjusted, we have precisely pointed out the consequences of this adjustment. We are not aware of any gross experimental or conceptual error in our analysis.

In the first stage of this work we have redetermined the absorption coefficient of hMDH with higher precision from its previous value. We then calculated an absolute value $M_2 = 80\,700 \pm 2500$ from SD, SE and LS experiments for a number of NaCl concentrations. With this value of M_2 and values of s and D, between 1 and 5 M-NaCl, values $(\partial\rho/\partial c_2)_\mu$ were calculated. Next we calculated values for ζ_1 by eqn. (**4**) assuming three values of $\bar{v}_2$. We could show (Fig. 1) that the water and salt binding parameters B_1 and B_3 were not strongly affected by this variation in $\bar{v}_2$. We demonstrated the reliability of $(\partial n/\partial c_2)_\mu$ values (Table 2) for LS determination of correct molar masses in non-aggregating protein solutions.

With the values of $(\partial\rho/\partial c_2)_\mu$ and ζ_1 (Table 1) we could now calculate values of electron and neutron density increments (Tables 3 and 4) and M_2 values from these experiments, for various values of ζ_1. We note that in both XS and NS analyses M_2, though fluctuating statistically, does not vary with salt concentration, which in itself is a significant observation, and average values can be defined (small-angle X-ray detectors have improved since these measurements were undertaken [6], yet the analysis of protein solutions at very high salt concentrations remains a non-trivial problem). The interesting observation in Tables 3 and 4 is that with increasing assumed value $\bar{v}_2$ both M_2 values from X-ray- and neutron-scattering approach, though they do not quite reach, the absolute value derived from light scattering and the ultracentrifuge. Yet, as $\bar{v}_2$ increases M_2 from XS increases, in contrast with a decrease in the case of NS. With even higher $\bar{v}_2$ we could match all M_2 values, but it is hardly reasonable to assume values above $\bar{v}_2 = 0.77$. An average value 0.770 ± 0.011 can also be estimated from an analysis by Calmettes *et al.* [9] who combined eqns. (**4**) and (11) to estimate $\bar{v}_2$. Their work was undertaken in KCl, the major cytoplasmic electrolyte of *Ha. marismortui*, and they could show in all their measurements (SD, SE, NS) that if ρ or ρ_N are taken as experimental variables, all data for hMDH in both NaCl and KCl superimposed, when expressed on a molar basis. We found no way of obtaining a match to all experimental data presented in this work by taking either 65 632 (the dimer sequence value) or 131 264 (the tetramer sequence value) for M_2.

There is no point, at this stage of the analysis, in pursuing the argument further. We are confident that this quandary will be solved in due time with the completion of density increments and other experiments in progress. We are grateful for the opportunity of presenting at this meeting an in depth analysis of the combined use of a number of complementary biophysical technologies, towards the solution of an important problem.

References

1. Eisenberg, H., Mevarech, M. & Zaccai, G. (1992) Adv. Protein Chem. **43**, in the press
2. Mevarech, M., Eisenberg, H. & Neumann, E. (1977) Biochemistry **16**, 3781–3785
3. Mevarech, M. & Neumann, E. (1977) Biochemistry **16**, 3786–3791
4. Pundak, S. & Eisenberg, H. (1981) Eur. J. Biochem. **118**, 463–470
5. Pundak, S., Aloni, H. & Eisenberg, H. (1981) Eur. J. Biochem. **118**, 471–477
6. Reich, M. H., Kam, Z. & Eisenberg, H. (1982) Biochemistry **21**, 5189–5195

7. Zaccai, G., Wachtel, E. & Eisenberg, H. (1986) J. Mol. Biol. **190**, 97–106
8. Zaccai, G., Bunick, G. J. & Eisenberg, H. (1986) J. Mol. Biol. **192**, 155–157
9. Calmettes, P., Eisenberg, H. & Zaccai, G. (1987) Biophys. Chem. **26**, 279–290
10. Zaccai, G., Cendrin, F., Haik, Y., Borochov, N. & Eisenberg, H. (1989) J. Mol. Biol. **208**, 491–500
11. Harel, M., Shoham, M., Frolow, F., Eisenberg, H., Mevarech, M., Yonath, A. & Sussman, J. L. (1988) J. Mol. Biol. **200**, 609–610
12. Fenselau, C. (1991) Annu. Rev. Biophys. Biophys. Chem. **20**, 205–220
13. Casassa, E. F. & Eisenberg, H. (1964) Adv. Protein Chem. **19**, 287–395
14. Eisenberg, H. (1976) Biological Macromolecules and Polyelectrolytes in Solution, Clarendon Press, Oxford
15. Mevarech, M. (1977) Ph.D. Thesis, Weizmann Institute of Science, Rehovot, Israel
16. Pundak, S. (1980) Ph.D. Thesis, Weizmann Institute of Science, Rehovot, Israel
17. Leicht, M. (1979) Ph.D. Thesis, Weizmann Institute of Science, Rehovot, Israel
18. Mevarech, M. & Werber, M. (1978) Arch. Biochem. Biophys. **187**, 447–456
19. Hase, T., Wakabayashi, S., Matsubara, H., Mevarech, M. & Werber, M. M. (1980) Biochim. Biophys. Acta **623**, 139–145
20. Voordouw, G., Kam, Z., Borochov, N. & Eisenberg, H. (1978) Biophys. Chem. **8**, 171–189
21. Chan, P. T., Ohmori, H., Tomizawa, J. & Lebowitz, J. (1985) J. Biol. Chem. **260**, 8925–8935
21a. Ebel, C., Guinet, F., Langowski, J., Urbanke, C., Gagnon, J. & Zaccai, G. (1992) J. Mol. Biol. **223**, 361–371
22. Baldacci, G., Guinet, F., Tillit, F., Zaccai, G. & de Recondo, A.-M. (1990) Nucleic Acids Res. **18**, 507–511
23. Cohen, G. & Eisenberg, H. (1966) Biopolymers **4**, 429–440
24. Zamyatnin, A. A. (1972) Prog. Biophys. Mol. Biol. **24**, 107–123
25. Chothia, C. (1975) Nature (London) **254**, 304–308
26. Reisler, E., Haik, Y. & Eisenberg, H. (1977) Biochemistry **16**, 197–203
27. Tardieu, A. E., Vachette, P., Gulik, A. & Lemaire, M. (1981) Biochemistry **20**, 4399–4406
28. Jolly, D. & Eisenberg, H. (1976) Biopolymers **15**, 61–95
29. Doty, P. & Geiduschek, E. P. (1953) in The Proteins (Neurath, H. & Bailey, K., eds.), pp. 402–405, Academic Press, New York
30. Chiang, R. (1966) in Polymer Handbook (Brandrup, J. & Immergut, E. H., eds.), Chapter IV, pp. 279–314, Interscience, New York
31. McMeekin, T. L., Wilensky, M. & Groves, M. L. (1962) Biochem. Biophys. Res. Commun. **7**, 151–156
32. Eisenberg, H., Josephs, R., Reisler, E. & Schellman, J. (1977) Biopolymers **16**, 2773–2783
33. Eisenberg, H. (1981) Q. Rev. Biophys. **14**, 141–172
34. Jacrot, B. & Zaccai, G. (1981) Biopolymers **20**, 2413–2426

Biochem. Soc. Symp. **58**, 127–133
Printed in Great Britain

Proteins from hyperthermophilic archaea: stability towards covalent modification of the peptide chain

Reinhard Hensel*†, Irmgard Jakob, Hugo Scheer‡ and Friedrich Lottspeich

Max-Planck-Institut für Biochemie, 8033 Martinsried, Germany

Introduction

One of the most striking features of the Archaea is their extraordinary thermophilic potential. Thus, hyperthermophiles with optimal growth temperatures above 80 °C are for the most part archaea and all organisms with growth temperatures above 100 °C isolated up to now belong exclusively to the domain of Archaea [1,2].

Life at these high growth temperatures requires provision by the cell of proteins which are stable and biologically active at these extreme temperatures. Although several archaeal proteins exhibiting astonishingly high thermostabilities and high temperature optima of activity have been characterized, only in a few cases are structural data available. The primary structures of only about a dozen proteins from archaeal hyperthermophiles have been analysed. In no case could the three-dimensional structure be resolved, which might have allowed insights into the construction principle of these proteins.

Thus, the question remained unanswered as to how the proteins from these hyperthermophiles are protected against covalent damage at temperatures which induce chemical modifications in 'normal' mesophilic proteins. As has been shown by several authors [3–8] these chemical modifications comprise deamidation of Asn

* To whom correspondence should be addressed.

† Present address: Universität GHS Essen, Universitätsstr. 2, 4300 Essen 1, Germany

‡ Present address: Botanisches Institut der Universität München, Menzingerstr. 67, 8000 München 19, Germany

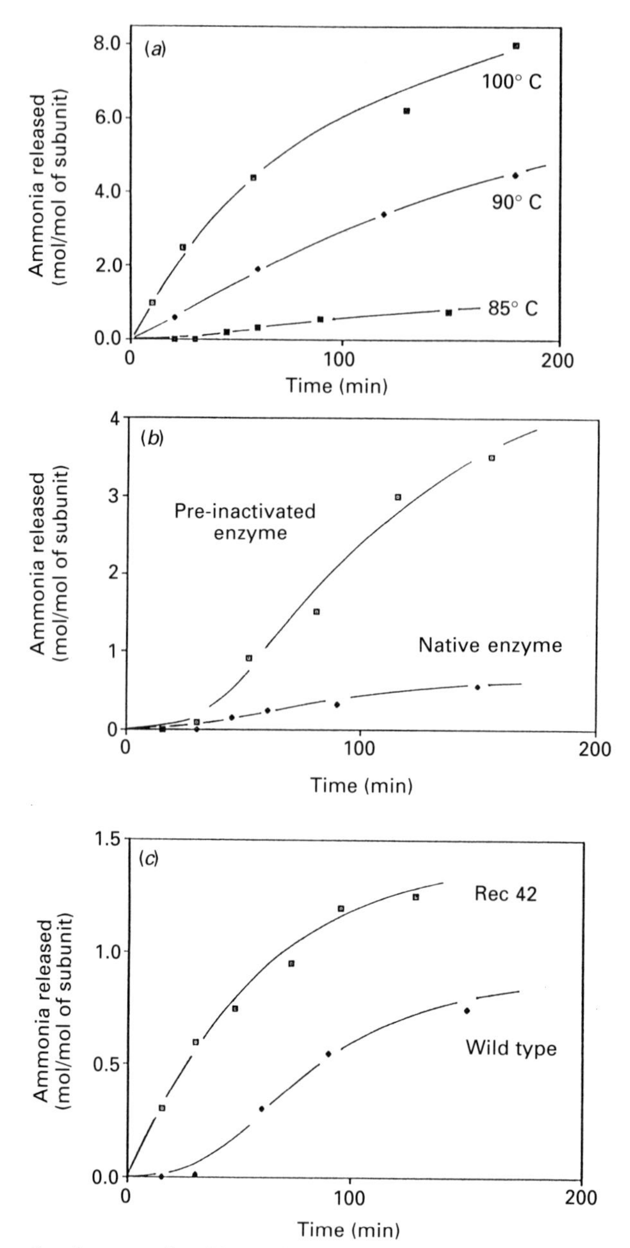

Fig. 1. Ammonia liberation kinetics of the GAPDHs from *M. fervidus* and *P. woesei*. The incubations were performed at pH 7.3 (10 mM-potassium phosphate buffer containing 150 mM-2-mercapto-ethanol; protein concentration: 2 mg/ml) in sealed glass capillaries under anaerobic atmosphere. Before the determination of ammonia by an amino acid analyser, the samples (70 μl) were cooled down immediately after incubation and mixed with 210 μl of 0.2 M-citrate buffer, pH 2.2. (*a*)

and Gln, hydrolysis of Asp-containing peptide bonds and Asn-Xaa bonds as well as destruction of cystine bonds.

Here we describe the susceptibility of archaeal hyperthermophilic proteins to thermogenic covalent modifications. The studies were performed with the glyceraldehyde-3-phosphate dehydrogenase (GAPDH) from *Methanothermus fervidus* (optimal growth temperature 83 °C [9]) and *Pyrococcus woesei* (optimal growth temperature 100 °C [10]) [11,12]. We have focused on deamidation and hydrolysis of the peptide bonds at high temperatures. To avoid oxidation reactions all incubations were performed under anaerobic conditions in the presence of reducing agents. Destruction of disulphide bonds was disregarded, since no indications for cystinyl cross-links in these proteins are available [12,13].

Results and discussion

Deamidation reactions at high temperatures

To investigate the susceptibility of the GAPDHs from the hyperthermophiles *M. fervidus* and *P. woesei* to deamidation reactions, ammonia liberation from the protein solutions (2.0 mg/ml) was followed at different temperatures. The proteins were incubated at low ionic strength (10 mM-potassium phosphate buffer, pH 7.3), at which the proteins show only a low stability as measured by inactivation kinetics [12,14]. Thus, at the respective optimal growth temperatures of the organisms the enzymes exhibit half-lives of inactivation of 60 min at 83 °C (GAPDH from *M. fervidus*) or of 44 min at 100 °C (GAPDH from *P. woesei*).

As shown in Fig. 1*a* and 1*b*, ammonia liberation can be observed with both enzymes. The GAPDH from *M. fervidus* exhibits, however, a higher susceptibility to deamidation than that from *P. woesei*; at 85 °C the *M. fervidus* enzyme already shows the same ammonia liberation rate as that obtained for the more stable *P. woesei* enzyme at 100 °C. Quite obviously, the enzyme proteins of the hyperthermophilic archaea are not resistant *per se* to chemical modification but must be protected against covalent damage.

Does the native conformation protect the Asn and Gln residues against deamidation?

To investigate whether the Asn and Gln residues are protected in the native conformation, the deamidation reaction was analysed in proteins with destabilized or disturbed conformation.

For this purpose the velocity of deamidation was determined with Rec 42, an enzyme mutant of the *M. fervidus* GAPDH. This mutant enzyme represents a hybrid enzyme constructed by recombination of the structural GAPDH genes from the thermophile *M. fervidus* and the mesophile *Methanobacterium bryantii* [14]. The hybrid GAPDH consists mainly of the *M. fervidus* GAPDH sequence but exhibits at its *C*-terminus a 42 residue fragment of the *Mb. bryantii* enzyme. This exchange resulted in

---

M. fervidus wild-type GAPDH incubated at different temperatures. (*b*) Native and pre-denaturated (see text) *P. woesei* GAPDH incubated at 100 °C. (*c*) *M. fervidus* wild-type and mutant (Rec 42) GAPDH incubated at 85 °C.

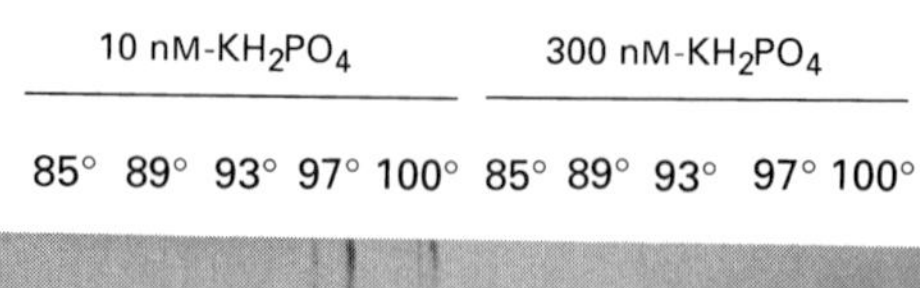

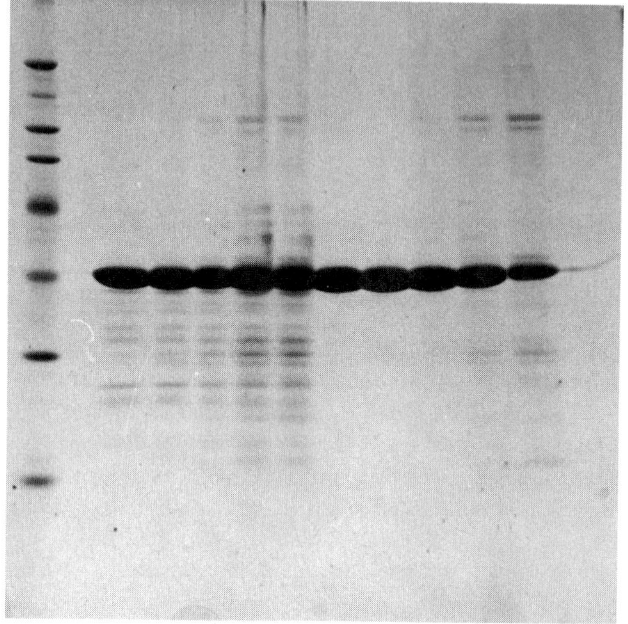

Fig. 2. Hydrolysis of peptide bonds. Electropherogram of *M. fervidus* GAPDH incubated for 1 h at different temperatures at low (10 mM-potassium phosphate buffer, pH 7.3 containing 150 mM-2-mercapto-ethanol) or at high ionic strength (300 mM-potassium phosphate buffer, pH 7.3 containing 150 mM-2-mercaptoethanol).

real substitution of only 10 residues [14]. No additional Asn or Gln residues were inserted in the recombinant structure. On the contrary, two Asn residues were substituted and the *C*-terminal Gln residue was eliminated by this recombination.

As documented previously [14], this rather minor exchange greatly destabilizes the protein structure (half-life of inactivation at 85 °C: 0.5 min (Rec 42) as opposed to 20 min determined for the wild-type GAPDH from *M. fervidus*). With the lower stability of the native conformation the susceptibility to deamidation reactions also increases, as shown in Fig. 1*c*. Thus, the retardation of ammonia evolution, characteristic for the wild-type enzyme from *M. fervidus*, disappeared in the case of Rec 42. Quite obviously, the conformational stability of the mutant enzyme is too low to retard the deamidation reaction.

The assumption that the network of non-covalent bonds in the native state protects the peptide chain against deamidation is also supported by deamidation experiments with the *P. woesei* GAPDH disrupted in its native conformation by

Fragment	Sequence	
	M. fervidus	*P. woesei*
1	G L S F N ▼ S L S N	Q V S F V S S S N
2	A I I P N ▼ P P K L	A I K P . S V T I
3	M H Q H N ▼ V M V E	M H V H S I M V E
4	V V S C N ▼ T T G L	V V S C N T T G L
5	K G P I N ▼ A I I P	R G P I N A I K P

Fig. 3. Cleavage sites of the non-enzymic hydrolysis of the *M. fervidus* GAPDH and comparison with homologous sequences of the *P. woesei* GAPDH. The positions of cleavage sites (marked by arrows and large type-size symbols) were deduced from the *N*-terminal amino acid sequence of the respective peptide fragments. Positions homologous to the cleavage sites in the *M. fervidus* GAPDH are underlined. Sequencing of the peptides was performed on a gas-phase sequencer (A470 from Applied Biosystems) after electroblotting on glass fibre sheets [18].

pretreatment with 8 M-guanidium chloride at 70 °C (Fig. 1*b*). This irreversibly denatured protein (after denaturation the denaturant was removed by dialysis to avoid interference with ammonia liberation) shows a significantly faster ammonia liberation than the native enzyme, indicating that in the disrupted conformation more residues are susceptible to covalent modification than in the native state. Nevertheless, as in the case of the pretreated enzyme a sigmoidal curvature can be observed, which accounts for a certain conformational stability of the wrongly refolded state.

The conclusion that the deamidation requires an unfolding of the chain seems to be plausible considering the reaction mechanism proposed by Clarke [4]. As outlined, the reaction starts with a nucleophilic attack of the peptide amino group on the amide carbon, forming a cyclic intermediate, a succinimide derivative. This reaction pathway requires that the ψ and χ angle assume values of $-120°$ and $120°$, which, however, is only possible in an unfolded state.

Thermogenic hydrolysis of the peptide bond

Further hints that conformational stability governs the susceptibility of hyperthermophilic proteins to chemical modifications can be deduced from studies on the thermogenic hydrolysis of peptide bonds in the GAPDHs from *M. fervidus* and *P. woesei*.

As documented in the left part of Fig. 2, incubation at or above 85 °C causes a fragmentation of the peptide chain of the *M. fervidus* GAPDH, however, only at low ionic strength, i.e. under non-stabilizing conditions. At a high phosphate concentration (right part of Fig. 2), known to stabilize proteins from this methanogen [15], the reaction is hindered.

From the sequences of the *N*-termini of the fragments the respective cleavage sites could be deduced. As shown in Fig. 3, hydrolysis occurs exclusively next to Asn residues. Obviously, the often described cleavage of the peptide bond at Asp residues takes place mainly at acidic pH and is not relevant at physiological (neutral) conditions.

The non-enzymic cleavage at Asn residues has already been described in proteins from mesophilic organisms [7,8]. Like the deamidation reaction, the cleavage also proceeds via a cyclic succinimide derivative, but in contrast with the deamidation reaction, the cyclization starts with a nucleophilic attack of the amide nitrogen on the carbonyl carbon of the peptide bond, thus leading to the cleavage.

As with the deamidation reaction [15a], since the cleavage reaction itself is favoured at high ionic strength, the hindrance of hydrolysis at high ionic strength, as observed in the case of the *M. fervidus* GAPDH, must be due to the extrinsic stabilization of the protein conformation.

Resistance of *P. woesei* GAPDH to non-enzymic hydrolysis: hints for deactivation of the weak links of the peptide chain by substitution or elimination of Asn residues

As one can expect from its higher resistance to deamidation, the GAPDH from *P. woesei* also shows a significantly lower susceptibility to hydrolysis of the peptide bonds. Thus, after incubation for 1 h at 100 °C no, or very few, hydrolysis products are visible in the respective electropherograms (not shown). We assume that, for the most part, the higher conformational stability of this protein causes its higher resistance to non-enzymic hydrolysis.

Additionally, we speculate that in proteins adapted to extremely high thermal conditions, 'hot spots' of chemical modification are avoided, especially in flexible structure elements.

In this regard we interpret the finding that the *P. woesei* GAPDH sequence lacks Asn residues at three positions, which are homologous to the 'fragile' positions in the *M. fervidus* sequence (Fig. 3).

At two 'fragile' positions, however, Asn residues are conserved in the *P. woesei* structure, probably for functional reasons. By analogy with the three-dimensional structure of the *Bacillus stearothermophilus* GAPDH [16], the conserved Asn residue in fragment 4 is the neighbour of the catalytically essential Cys residue, whereas the other conserved Asn residue in fragment 5 is located in the functionally important S-loop. Obviously, these residues are protected by the rigid conformation of the *P. woesei* GAPDH.

Because of their chemical lability we would expect that Asn residues are generally reduced in number in proteins adapted to the higher temperature range (above 80 °C). Comparing the GAPDH and 3-phosphoglycerate kinase sequences from mesophilic and thermophilic archaea [12,17] the expected tendency can be confirmed. Similar trends are also visible in bacterial proteins; strikingly low Asn content was found in enzyme proteins from *Thermus* strains with upper growth temperatures around 85 °C indicating that the requirements for the construction of proteins for the higher temperature range are similar in both domains.

The work was supported by grants from the Deutsche Forschungsgemeinschaft and the Fonds der Chemischen Industrie.

References

1. Stetter, K. O., Fiala, G., Huber, R. & Segerer, A. (1990) FEMS Microbiol. Rev. **75**, 117–124
2. Woese, C. R., Kandler, O. & Wheelis, M. L. (1990) Proc. Natl. Acad. Sci. U.S.A. **87**, 4576–4579
3. Inglis, A. S. (1983) Methods Enzymol. **91**, 324–332
4. Clarke, S. (1985) Annu. Rev. Biochem. **54**, 479–506
5. Ahern, T. J. & Klibanov, A. M. (1985) Science **228**, 1280–1284
6. Zale, S. E. & Klibanov, A. M. (1986) Biochemistry **25**, 5432–5444
7. Geiger, T. & Clarke, S. (1987) J. Biol. Chem. **262**, 785–794
8. Voorter, C. E. M., de Haard-Hoekman, W. A., van den Oetelaar, P. J. M., Bloemendal, H. & de Jong, W. W. (1988) J. Biol. Chem. **263**, 19020–19023
9. Stetter, K. O., Thomm, M., Winter, J., Wildgruber, G., Huber, H., Zillig, W., Janekovic, D., König, H., Palm, P. & Wunderl, S. (1981) Zentralbl. Bakteriol. Mikrobiol. Hyg. Abt. 1 Orig. C **2**, 166–178
10. Zillig, W., Holz, I., Klenk, H.-P., Trent, J., Wunderl, S., Janekovic, D., Imsel, E. & Haas, B. (1987) Syst. Appl. Microbiol. **9**, 62–70
11. Fabry, S. & Hensel, R. (1987) Eur. J. Biochem. **165**, 147–155
12. Zwickl, P., Fabry, S., Bogedain, C., Haas, A. & Hensel, R. (1990) J. Bacteriol. **172**, 4329–4338
13. Fabry, S. & Hensel, R. (1988) Gene **64**, 189–197
14. Biro, J., Fabry, S., Dietmaier, W., Bogedain, C. & Hensel, R. (1990) FEBS Lett. **275**, 130–134
15. Hensel, R. & König, H. (1988) FEMS Microbiol. Lett. **49**, 75–79

15a. Robinson, A. B. & Rudd, C. J. (1974) in Current Topics in Cellular Regulation (Horecker, B. L. & Stadtman, E. R., eds.), Vol. 8, pp. 247–295, Academic Press, New York

16. Biesecker, G., Harris, J., Thierry, J. C., Walker, J. E. & Wonacott, A. J. (1977) Nature (London) **266**, 328–333
17. Fabry, S., Heppner, P., Dietmaier, W. & Hensel, R. (1990) Gene **91**, 19–25
18. Eckerskorn, C., Mewes, W., Goretzki, H. & Lottspeich, F. (1988) Eur. J. Biochem. **176**, 509–519

Biochem. Soc. Symp. **58**, 135–147
Printed in Great Britain

Biotechnological potential of halobacteria

Francisco Rodriguez-Valera

Departmento de Genética Molecular y Microbiología, Universidad de Alicante, Campus de San Juan, Apartado 374, 03080 Alicante, Spain

Synopsis

The extremely halophilic archaebacteria (halobacteria) became an early focus of scientific interest owing to their role in salted food deterioration. In more recent times their peculiar physiology involving extreme adaptation to the salt environment and other unique features have allowed the development of other applied interests. Their similarities to eukaryotic cells at the level of cell division justifies their use in the prescreening for anti-cancer drugs, and some of their antigens could be used for cancer diagnosis. Their unique retinal proteins can be used as light-biosensors and the use of the purple membrane (pm) as reversible holographic medium has already been developed. Halobacterial enzymes are an extremely tough raw material for enzyme technology, particularly for applications in which the reaction mixture has very low water activity. Thanks to their peculiar lipids and to the production of polysaccharides by some halobacteria, their cultures could be used for enhanced oil recovery. Some halobacteria are excellent producers of industrially interesting biopolymers. The use of halobacteria as producers of polyhydroxyalkanoates, biological polyesters such as poly-3-hydroxybutyrate, with the properties of biodegradable thermoplastics, is being considered.

Introduction

'The great diversity among bacterial and eukaryotic cells with respect to their chemical composition, metabolic and biosynthetic abilities, and genetic make up is the horn of plenty from which biotechnologists obtain their play material'. This quotation from Kandler [1] makes what, in my opinion, is the main argument in favour of the high potential of archaebacteria as subjects of biotechnological interest. Their vast reserve of diversity with regard to basic cell properties guarantees a substantial contribution of this unique group of micro-organisms to the future of biotechnology. Therefore, I will start this review by considering the basic aspects of halobacteria that mark their uniqueness and hence their specific contribution to the diversity of global life.

Singularities of halobacteria

The aerobic halophilic archaebacteria are now included in the order Halobacteriales [2], including one family and six genera. They are rods, cocci or pleomorphic cells with cell walls made up of either glycoprotein S-layers (rods and some pleomorphic shapes) [3,4] or a sulphate-rich heteropolysaccharide (cocci) [5]. An intermediate type of cell wall with both types of polymers could also occur in some pleomorphic halobacteria (J. Anton, I. Meseguer & F. Rodriguez-Valera, unpublished work). Metabolically speaking most halobacteria show a typical aerobic chemoorganotrophic metabolism (some are also able to grow anaerobically, fermenting amino acids or reducing nitrate) strongly reminiscent of that found in many aerobic eubacteria. That means that they are archaebacteria of 'modern' metabolism that can grow with high yields and under conditions identical to those used for growing aerobic eubacteria. However, under certain conditions some halobacteria can exhibit a peculiar phototrophic metabolism based on retinal proteins. The purple membrane produced by some halobacteria, consisting exclusively of bacteriorhodopsin (BR) and lipids, acts as a light-driven proton pump generating energy for the cell. Although pure autotrophic growth has never been demonstrated for halobacteria, CO_2 fixation by anaplerotic reactions has been shown for BR-containing species [6]. Ribulose bisphosphate carboxylase (RuBP) activity (the key enzyme of the Calvin cycle) and RuBP-dependent CO_2 fixation has been detected in cell-free extracts, albeit in species that do not have BR [7]. Therefore, photoautotropic growth could not occur in those organisms, instead some evidence indicates that the reduction power could be derived from chemolithotrophic reactions (H_2 oxidation).

The environment in which halobacteria thrive is in itself a pecularity; they can be found in large numbers in NaCl-saturated waters as virtually exclusive inhabitants [8,9]. The capacity to grow in saturated NaCl is very unusual, other halophilic organisms can approach the saturation levels but scarcely grow when nearing this limit (some strains of the photosynthetic eubacteria *Ectothiorhodospira halophila* are the sole exception of which I am aware). Halobacteria grow readily at saturating concentrations, provided that oxygen is available (for example on solid media). But it is even more remarkable that saturated NaCl is their principal habitat, in which the largest halobacterial populations are found. The dense biomass in the saturated ponds of solar salterns is, with all probability, made up exclusively of halobacteria [9]. In other words it can be said that halobacteria are the most halophilic halophiles, in the same way as the hyperthermophilic archaebacteria are the most thermophilic organisms known. Profound extremophily appears to be a common feature of archaebacteria.

The halobacteria are certainly also the organisms that maintain the highest concentrations of inorganic ions in their cytoplasm, with roughly 5 M intracellular K^+ when growing optimally [10], while all other known halophiles (with the exception of some groups of moderately halophilic anaerobic eubacteria [11]) accumulate compatible organic solutes that protect their enzymes and exclude most ions from their cytoplasm. This means that their whole internal machinery works exposed to extremely high ionic strength. Halobacterial proteins have unusually high proportions of acidic amino acids and low proportions of polar and basic residues [12]. This could help to maintain an adequate balance of hydrophobic–electrostatic interactions [13], as well as a proper level of hydration [14], in the high-salt environment.

Halobacteria and food

The first scientific interest awakened by halobacteria derived from their undesired growth on salted food, mostly fish. When this commodity is stored in warm and humid conditions a red patina, known in the trade as the 'pink', develops occasionally, corresponding to the growth of halobacteria of either *Halobacterium* or *Halococcus* species [15]. Since both genera are actively proteolytic, considerable degradation of the material can occur. That, together with a reported unpalatability (at least to Western taste) of halobacterial biomass, leads to the irreversible loss of the product. In present days the problem of deterioration of salted food has been greatly reduced by refrigeration of the finished product and lower temperatures during processing. Still, a cheap and safe (for consumers) method of eliminating the halobacteria present in sea salt or preventing their growth on the food would certainly be welcomed by this industry. On the other hand, halobacteria able to grow at 4 °C, albeit at an extremely low rate, have been described recently [16]. Although these organisms have been isolated from Antarctica, and it is highly probable that they are not present in sea salt, it is not unthinkable that they could eventually reach salted food processing plants.

Some inhibitors of halobacterial growth that have been described recently are of prospective interest. Some bile salts, taurocholate for rods and deoxycholate for cocci for example, have shown a high efficiency for lysing halobacterial cells at extremely low concentrations (16–45 mg/l) [17]. Mixed with salt, addition of these compounds could be an effective system of preventing halobacterial development in the food.

On the other hand halobacterial growth in food is not always undesired. Fish sauce, a widely used food condiment in southeast Asia (known as Nam Pla in Thailand), seems to be, at least partially, the result of the proteolytic activity of halobacteria on fish proteins [18]. In fact, considerable amounts of these microorganisms could be present in the final product constituting, to my knowledge, the sole example of arachaebacteria consumed as food. There is no information about the effects of ingestion of halobacterial biomass in normal diet; considering the peculiar lipid constitutents (both polar and non-polar) of halobacteria and archaebacteria in general, it is interesting to consider what effects on human health the consumption of such compounds on a daily basis could have.

Halobacteria and cancer research

In spite of the apparent unrelatedness of both concepts, two independent approaches have been described in which halobacteria can help in cancer research. Both methods take advantage of the remarkable similarities that exist at the molecular biology level between halobacteria and eukaryotic cells.

Microbial prescreens have played an important role in the discovery of anti-cancer agents [19]. However, eubacterial sensitivity is widely different from that of eukaryotic cells. On the other hand, yeasts present several permeability problems due to their complex envelopes. By contrast halobacteria have been shown to share sensitivity patterns with at least two major targets of cytostatic drugs: DNA topoisomerase II and cytoskeleton components (actin and tubulin) [20]. The first is based on a similarity of the corresponding enzymes. The sensitivity to cytoskeletal targeted drugs is more obscure; however, the existence of actin-like molecules in archaebacteria has been proposed already for *Thermoplasma* [21]. The existence in halobacteria of sequences homologous to the yeast actin gene and of proteins cross-reacting with antibodies produced against chicken actin and tubulin have been shown [20]. In fact, there is fragmentary ultrastructural evidence showing intracellular organelles in halobacteria, generally associated with the plasma membrane, consisting of bundles of hollow tubes that could well correspond to actin-like proteins [23,24] presumably implicated in dynamic cell processes such as cell division. Other components of the cell duplication machinery of halobacteria, such as the DNA polymerase, have also shown a similar sensitivity pattern to eukaryotic DNA polymerases [24]. Considering the uncomplicated cultivation of halobacteria and their simple cell wall, their use as microbial pre-screens could very well expedite the screening for new anti-cancer drugs [20].

Another aspect of halobacteria related to cancer research, in this case the diagnosis of certain types of cancer, is the remarkable similarity demonstrated between the *c-myc* gene product and an 84 kDa protein isolated from cell extracts of *Halobacterium halobium*. This protein cross-reacts with antibodies raised against the human *c-myc* gene product and can be used for immunological tests devised to detect the levels of these antibodies in patients' serum, which is clinically relevant [25]. Surprisingly the positive reaction against the halobacterial protein had higher correlation with colorectal cancer than the human *c-myc* protein produced in *Escherichia coli* by genetic engineering [27]. On the other hand, halobacterial α-like DNA polymerase cross-reacts with antibodies against the *v-myc* oncogene product indicating a possible relationship between *myc* protein and the halobacterial enzyme [27].

Retinal proteins

The retinal proteins of halobacteria represent one of the most peculiar characteristics of these organisms. Since the discovery of BR in the early 1970s several practical applications of this biological light-driven proton pump have been scrutinized, from desalinization of seawater [28] and photocurrent generation [29] to optical switching elements of 'biolelectronics' [30]. I will consider briefly some of the more recent and promising developments in this rapidly evolving field related to the use of BR as an optical recording material and as a molecular computing element.

For those applications BR offers several benefits. Isolated BR or PM is very stable and easy to immobilize on solid substrates, such as glass plates, or to embed in polymers, maintaining their photochemical properties over very long periods. Proton pumping in a model membrane for at least one year has been reported [31]. Their photochemical reactions are self-regenerative, and their photoelectric signal is extremely reproducible [30]. Moreover, the molecular biology of this photoreceptor is very well known, to the extent of identifying the role of specific amino acids in the properties of the macromolecule regarding the light-response, proton pumping and photocycle.

One application already in an advanced state of development is the use of PM films as holographic media [32]. As such, BR can work in a similar way to the classical photographic film where the interference patterns, that constitute the holographic recording, are stored as black or white dots on the silver halide film. A BR molecule can function similarly with two alternatives—ground purple B state or yellow M intermediate. The interference patterns are registered as purple or yellow areas, depending on the intermediary of the photocycle corresponding to each BR molecule on the immobilized PM. Two alternative recording mechanisms have been proposed [33]. One mechanism is based on the photoactivity of the BR ground state after illumination with a wide band of wavelengths (500–680 nm), transition B → M. An alternative system is the transition M → B of a PM film previously populated with M state by illumination in the spectral range of the B state and then illuminated with light of 400–450 nm (M-state absorption maximum). M-type holograms offer certain advantages when a mutant BR with longer M-state lifetime is used instead of the wild-type BR [32]. This exemplifies the possibilities of applying genetic techniques, particularly genetic engineering, to the system. Whatever mechanism is used, a PM film does not require developing and is reversible, i.e. a PM matrix can be recorded several times.

These two properties make PM film particularly suitable for some specific applications of holography. One is the use of holography as computer memory, with several advantages over other optical storage materials such as the CD-ROM (Compact Disc-Read Only Memory) [34]. Not only can the amount of information stored be much higher, but also all the information is processed simultaneously. This would be especially suitable for parallel processing. Computers capable of parallel processing, i.e. able to process more than one information source (ideally many) simultaneously, are one of the most promising lines of development of computer science. With semiconductor-based computer memories of the type used presently, parallel processing requires complex electronic set-ups with severe limitations. Holographic memories would simplify parallel processing because all the information stored will be available simultaneously. Holography allows parallel storage and processing of information as a built-in property of the system. On the other hand a PM memory would be reversible, i.e. can be recorded several times changing or modifying the information stored—a property that very few holographic materials possess—and with the essential advantage of being an easy to produce material with high homogeneity and low cost. The use of PM holograms for optical pattern recognition in real time, a technique very helpful, for instance, in processing weather satellite information or robot vision, has been the first achievement in this line of work (N. Hampp, R. Thoma, D. Oesterhelt & C. Braüche, unpublished work).

Another potential application of BR in computers is as a computing element. Silicon-based microchips are rapidly reaching a ceiling in miniaturization and speed due to physical limitations [35]. This fact is stimulating research to look for substrates more prone to reduction, such as molecular or even atomic switches [36]. 'Biochips', where proteins or other biomolecules are the computing elements have been proposed as an alternative to electronic chips [35]. BR is an interesting material with which to develop an optical biomolecular computer where the two alternative states of the photocyle act as an optical switch analogous to the conducting/non-conducting states in a semiconductor.

The increasing number of halobacterial retinal proteins known, together with the possibility of their genetic manipulation, opens a wide field of opportunities for developing new technologies as exemplified by the PM films.

Enzymes

Protein fragility is one of the major drawbacks of enzyme technology. Therefore, the capability of certain enymes to work in stressful environments is considered an essential feature for several industrial applications [37]. In this sense, halobacterial enzymes have proven to be remarkably sturdy, and not only in regard to high salinity stress. Certainly, the intracellular enzymes of these organisms are the most halophilic and halotolerant [38] (the anaerobic eubacteria that accumulate salts intracellularly maintain much lower concentrations than the halobacteria [39]). Besides this, their extracellular enzymes are among the most halophilic produced by any organism. In addition most halobacterial enzymes are considerably thermophilic [40,41] and remain stable at room temperature over very long periods. All these features are of potential interest for enzyme technology.

Enzymic reactions carried out in organic solvents can be particularly favoured by the use of halobacterial enzymes. The long-standing belief that enzyme activity is limited to water-based environments has been proven erroneous [42]. In general terms, enzyme catalysis in organic solvents offers benefits such as greatly enhanced stability, alterations in substrate and enantiomeric specificities, molecular memory (activation by a pretreatment) and improved product yield [42]. One of the major problems of maintaining the enzymic activity in high proportions of organic solvents is the low water activity, particularly with the more hydrophilic ones [43]. Apparently, the more hydrophilic non-polar solvents tend to strip the essential hydration layer from proteins [42], but halobacterial enzymes can work in conditions of very low water activity and are very efficient in maintaining their hdyration layer. Therefore it is predictable that these enzymes could be most appropriate for use in these processes where the reaction is carried out in the presence of considerable amounts of organic solvents or in general where low water activity is found [44].

Another interesting aspect of halobacterial enzymes is as a source of basic knowledge, that can also be of application. The most powerful recent development in enzymology is protein engineering [45]; that is the redesign of enzymes by site-directed mutagenesis. In a not too distant future it is predictable that enzymes will be tailored by genetic engineering to catalyse specific reactions under specific conditions. In this sense the basis for the haloresistance of halobacterial enzymes in terms of amino acid sequence and structure would be essential to predict the salt resistance and water activity requirement of these 'artificial' enzymes [44].

Halobacteria and oil

It is an accepted fact nowadays that hypersaline environments have played an essential role in preserving large amounts of biogenic organic matter throughout geological eras, contributing to the formation of some of the largest petroleum reservoirs on Earth [46]. Although the evaporitic environments that have contributed more significantly are the ones designated as the carbonate domain by geologists, and therefore with lower salinities than those required for halobacterial development [47], the peculiar cell composition of these micro-organisms can be used as a biomarker of ancient hypersaline-evaporitic environments and indirectly as a help in oil prospection. Certain lipids that are characteristic of halobacteria and not found in other archaebacteria are promising in this sense. Ether-linked lipids do have a long geological survival time [48].

Another potential use of halobacteria related to petroleum is their use in enhanced (or tertiary) oil recovery (EOR). This is basically a forced extraction of the crude oil retained in the deposit by injection of water in a separate perforation into the reservoir. This water displaces the oil, retained in strata after conventional recovery methods have been utilized, the oil is then pushed to the surface through the oil well. The process is much more efficient if the properties of water are modified, increasing its viscosity and decreasing its surface tension [49]. Halobacterial lipids have surfactant properties with a hydrophile–lipophile balance in the optimum range for enhanced oil recovery [50]. In addition, it is extremely frequent that the conditions in the oil deposit are saline due to the typical halite strata very often associated with the reservoir. Therefore the use of a saline-resistant surfactant is very convenient. In addition some halobacteria produce exopolysaccharides that can improve the rheological properties of water for EOR (see next section).

Polysaccharides

Several polysaccharides of microbial origin are of applied interest, in many cases because of their ability to vary the rheological properties of water (viscosity) at low concentrations [51]. Anionic polysaccharides are particularly useful, and are utilized as gelling agents, stabilizers, thickeners and emulsifiers in the food, paint, photographic and pharmaceutical industries among others. More sophisticated applications, such as the use of sulphated polysaccharides as anti-viral [52] and anti-tumoral [53] agents, are also under study. Although anionic polysaccharides are produced by a wide diversity of microorganisms, at present commercial exploitation is restricted to algae (for example agar-agar or carrageen) or eubacteria (for example xanthan gum). However, alternative sources are being actively sought to fulfil the increasing demand and diversify the range of these polymers on offer [54]. Some archaebacteria also produce extracellular polysaccharides. Specifically, *Methanosarcina* [55] and some species of *Haloferax* [56] and *Haloarcula* (J. Anton, I. Meseguer & F. Rodriguez-Valera, unpublished work) produce exopolysaccharides. In the case of the polysaccharide produced by *Haloferax mediterranei* the polymer has been partially characterized, proving to be a highly sulphated and acidic heteropolysaccharide [56]. The composition of the polymer varies depending on the culture medium; when grown in glucose medium the polymer contains glucose and galactose (1:11) as major neutral sugars but when grown in complex medium (yeast extract) mannose is the major component [57]. In both cases sulphate amounts to almost 6% of the polymer dry weight. The rheological properties of these polymer/water mixtures are very good (high viscosity at low concentrations) and extremely resistant to all kinds of stresses, including salinity as could be expected, but also temperature and pH [56]. That makes the polymer an interesting candidate for uses where a tough thickening agent is required. One of the best examples of these requirements is the application of polysaccharides for enhanced oil recovery. In EOR the addition of polymers to the injection water is recommended to improve the sweep efficiency [58]. The water injected in a petroleum reservoir is submitted to extremes of temperature, pressure and very often salinity [59]. Therefore a culture of a polysaccharide-producing halobacteria could be used as a whole for improving the recovery of crude oil, the extracellular polysaccharide would contribute to the adequate viscosity of the mixture while the lipids liberated from lysed cells would act also as surfactants to improve the oil carrying properties of the water.

Halobacteria as producers of bioplastics

Recently, another type of bacterial polymer has aroused industrial interest, the polyhydroxyalkanoates (PHAs), a heterogeneous family of polyesters utilized as carbon storage material that can be used as thermoplastics. Halobacteria may well prove expedient as producers of these polymers with the additional advantage represented by their easy cultivation. I will introduce first some of the characteristics and applications of PHAs.

PHAs or rather one type of PHA has been known for a long time, poly(3-hydroxybutyrate) (PHB) is the most common (at least in laboratory conditions) carbon reserve material accumulated by eubacteria [60]. When such micro-organisms are confronted with situations of abundant carbon supply but limitation of other essential nutrients such as phosphorus or nitrogen, the acetyl-CoA molecules that normally enter the tricarboxylic acid cycle are polymerized, passing through a four carbon (3-hydroxybutyrate, HB) intermediary, into a growing chain of PHB. The polymer, that acts as a carbon and electron sink, is accumulated intracellular in highly refractile granules of 0.1–0.5 μm diameter surrounded by a thin (2 nm) membrane composed of lipid and protein [61,62]. Whenever the stored carbon is required again, the granules are depolymerized generating acetyl-CoA that readily enters the catabolic and biosynthetic pathways. The biochemistry and regulation of the polymer synthesis and degradation are well known [62]. Until the 1970s it was thought that the PHB homopolymer was the only form accumulated by bacteria. Then a co-polymer containing two types of monomers, the usual hydroxybutyrate and the five carbon 3-hydroxyacid or 3-hydroxyvalerate (HV), was found in sewage sludge [63]. These co-polymers receive the general denomination of PHAs and can in fact be extremely diverse. The polymerase system present in some micro-organisms can incorporate a whole range of monomers that in the case of *Pseudomonas oleovorans* includes chain lengths of up to 12 carbon atoms as well as branched or cyclic residues [64]. In summary, a diverse set of polyesters can be synthesized by bacteria by selecting the appropriate organism and/or precursors. Additionally, the genes required for the synthesis of the polymer have been cloned, widening the possibilities still further [65–69].

The biotechnological interest of PHAs derive from their properties as thermoplastics. The PHB homopolymer has general properties similar to polypropylene [70]. As such, the material presents several disadvantages as a commodity plastic, being too rigid and brittle, and inadequate for thermal processing. However, these properties change in a wide range of co-polymers [70]. Co-polymers of HB and HV have much better properties with less stiffness, more strength and lower melting points. Since the amount of HV can be varied by simply modifying the composition of the culture medium, a whole range of the above mentioned properties can be obtained, from brittle polymers resembling unplasticized polyvinylchloride or polystyrene with the lower range of HV co-polymerization, to soft and tough ones similar to polyethylene with over 20% of HV monomers [71]. Other types of PHA show different properties, for example, co-polymers of 3-hydroxybutyrate and 4-hydroxybutyrate possess even lower crystallinity, producing extremely elastic polymers [72]. Summarizing, bacterial PHAs can be tailored to suit a wide range of applications currently filled by petrochemical plastics. In addition PHAs have further advantages; they are biodegradable and biocompatible (are tolerated and degraded in animal tissues). The first property permits their use as disposable plastics without creating the severe environmental problems that arise from massive discharge of petrochemical non-degradable plastics [64]. Biocompatibility can be utilized for several pharmaceutical and clinical applications, from retarded drug release [73] to surgical suture and bone replacements [70]. Other potential applications of PHAs lie in their chemical stereospecificity since the polymerase system incorporates only D(-) monomers to the growing chain. That can be used either as a system of purification of this type of stereoisomers or as a source of building blocks for the organic synthesis of fine chemicals such as insect pheromones that frequently correspond to this configuration in their active form [64].

The technical development and exploitation of HB–HV co-polymers as a commodity plastic has been carried out by the British corporation ICI. The polymer is synthesized by the soil eubacterium *Alicaligenes eutrophus* grown in standard fermenters in fed batch culture with glucose as carbon and energy source [71,74]. The different grades of HB–HV co-polymers are obtained by supplying different amounts of propionic acid to the medium and are commercialized under the trade name Biopol. The material can be processed using conventional methods to produce bottles, moulding, fibre and films. Biopol is already being used for cosmetic bottles. However, the present cost of the product (25–30 times the cost of polyethylene) prevents a more extensive market penetration [75]. Besides, with the conventional fermentation technology utilized, increasing the production of Biopol to supply a potential market of hundreds of thousands of tonnes per year would require a huge investment in the production plant.

It is in this context that halobacteria can be utilized with advantage. Some species accumulate considerable amounts of PHA [76]. In fact *H. mediterranei* has a number of advantages as producer when compared with. *A. eutrophus*. The amounts of PHA accumulated and the yield in relation to the carbon source are similar, but *H. mediterranei* can use starch as carbon source, a much cheaper substrate than glucose [77]. Moreover, this organism produces co-polymers of HB and HV even if no precursor is added into the medium (J. García Lillo & F. Rodriguez-Valera, unpublished work). With only starch as carbon source a minimum of 9% of HV subunits are incorporated in the polymer, which roughly corresponds with the lower limit that allows a good thermal processing. Therefore the cost of precursor can be reduced considerably. The fragility of halobacterial cells exposed to low salt concentrations allows the development of very simplified recovery processes.

Finally, an advantage that requires special consideration is the easiness of cultivation. In has been known for a long time that batches of up to 100 litres of halobacterial cultures can be grown in extremely simplified conditions [78]. In fact working with *Haloferax* that includes the halobacteria of fastest growth rates and widest physiological versatilities it is feasible to devise conditions in which there is virtually no possible contamination. In our laboratory we have maintained continuous cultures of *H. mediterranei* running for 3 months, with minimal sterility precautions. That allows the design of production facilities of vast dimensions that would require relatively little investment, resembling chemical reactors more than classical fermenters, and the process can be carried out in continuous form, optimizing production parameters far beyond the levels reached in batch processes. The idea of using extremophiles for contamination-free industrial microbiology is not new, and was already contemplated in the case of thermophiles [79]. For standard thermophilic processes it does not seem to be feasible [80], but it has never been attempted using hyperextremophiles such as archaebacteria often are.

Conclusions

Exploitation of the considerable contribution of halobacteria to the diversity of life is just beginning. Immobilized PM is already available for optical applications in basic, applied or pre-industrial research and shows strong indications of becoming a speciality product in the very near future. A pilot-scale plant for PHA production by halobacteria is in an advanced state of design and could be in operation in the near future. Diagnostic kits using halobacterial antigens could be used for diagnosis of certain types of cancer. However, the expectations for future developments are much higher. I hope that this review may help to stimulate further work on applications of halobacteria. Probably, most of such applications have not been contemplated in this review and are awaiting more basic knowledge as well as more scientists and technologists looking for them.

Part of the work from my laboratory discussed in this review was supported by grants PBT 86/0011, PTR89-0003 and BIO90-0475 from the CICYT (Comisión Interministerial de Ciencia y Tecnologia) of the Spanish Government. I wish to thank D. Oesterhelt for providing preprints of several papers and J. M. Iñesta, J. Grimalt and P. Berebel for valuable advice regarding specific parts of the manuscript. Secretarial assistance by K. Hernandez is gratefully acknowledged.

References

1. Kandler, O. (1984) Proc. 3rd Eur. Congr. on Biotechnol. Symp. München
2. Grant, W. D. & Larsen, H. (1989) in Bergey's Manual of Systematic Bacteriology, (Staley, J. T., Bryant, M. P., Pfennig, N. & Holt, J. G., eds.), vol. 3. pp. 2216–2219, Williams & Wilkins, Baltimore
3. Mescher, M. F. & Strominger J. L. (1976) Proc. Natl. Acad. Sci. U.S.A. **73**, 2687–2691
4. Kessel, M., Wildhaber, I., Cohen, S. & Baumeister, W. (1988) EMBO J. **7**, 1549–1554
5. Schleifer, K. H., Steber, J. & Mayer, H. (1982) in Archaebacteria (Kandler, O., ed.), pp. 171–178, Gustav Fischer Verlag, Stuttgart
6. Javor, B. J. (1988) Arch. Microbiol. **149**, 433–440
7. Altekar, W. & Rajagopalan, R. (1990) Arch. Microbiol. **153**, 169–174
8. Rodriguez-Valera, F., Ruiz-Berraquero, F. & Ramos-Cormenzana, A. (1981) Microbial Ecol. **7**, 235–243
9. Rodriquez-Valra, F., Ventosa, A., Juez, G. & Imhoff, J. F. (1985) Microbial Ecol. **11**, 107–115
10. Christian, J. H. B. & Waltho, J. A. (1962) Biochim. Biophys. Acta **65**, 506–508
11. Rengipat, S., Lowe, S. W. & Zeikus, J. G. (1988) J. Bacteriol. **170**, 3065–3071
12. Reistad, R. (1970) Arch. Mikrobiol. **71**, 353–360
13. Lanyi, J. K. (1974) Bacteriol. Rev. **38**, 272–290
14. Werber, M. W., Sussman, J. L. & Eisenberg, H. (1986) FEMS Microbiol. Rev. **39**, 129–135
15. Shewan, J. M. (1971) J. Appl. Bacteriol. **34**, 299–315
16. Franzmann, P. D., Stackerbrandt, E., Sanderson, K., Volkman, J. K., Cameron, D. E., Stevenson, P. L., McMeekin, T. A. & Burton, H. R. (1988) Syst. Appl. Microbiol. **11**, 20–27
17. Kamekura, M. & Seno, Y. (1991) in General and Applied Aspects of Halophilic Microorganisms (Rodriguez-Valera, F., ed.), pp. 395–365, Plenum Press, New York
18. Thongthai, C. & Suntinanalert, P. (1991) in General and Applied Aspects of Halophilic Microorganisms (Rodriguez-Valera, F., ed.), pp. 381–388, Plenum Press, New York
19. White, R. J. (1982) Ann. Rev. Microbiol. **36**, 415–433
20. Sioud, M., Baldarchi, G., Forterre, P. & de Recondo, A.-M. (1987) Eur. J. Biochem. **169**, 231–235
21. Searcy, D. G., Stein, D. B. & Searcy, K. B. (1980) Ann. N.Y. Acad. Sci. **361**, 312–323
22. Cho, K. Y., Doy, C. H. & Mercer, E. H. (1967) J. Bacteriol. **94**, 196–201
23. Robertson, J. D., Schreil, W. & Reedy, M. (1982) J. Ultrastruct. Res. **80**, 148–162
24. Forterre, P., Elie, C. & Kohiyama, M. (1984) J. Bacteriol. **159**, 800–802
25. Ben-Mahrez, K., Thierry, D., Sorokine, I., Danna-Muller, A. & Kohiyama, M. (1988) Br. J. Cancer **57**, 529–534
26. Ben-Mahrez, K., Sorokine, I., Thierry, D., Kawasumi, T., Ishii, S., Salmon, R. & Kohiyama, M. (1991) in General and Applied Aspects of Halophilic Microorganisms (Rodriguez-Valera, F., ed.), pp. 367–372, Plenum Press, New York
27. Sorokine, I., Ben-Mahrez, K., Nakayama, M. & Kohiyama, M. (1991) in General and Applied Aspects of Halophilic Microorganisms (Rodriguez-Valera, F., ed.), pp. 313–319, Plenum Press, New York

28. Prentis, S. (1981) New Sci. **101**, 159–163
29. Singh, K. & Caplan, S. R. (1980) Trends Biochem. Sci. **5**, 62–64
30. Hong, F. T. (1986) BioSystems **19**, 223–236
31. König, H. (1988) in Biotechnology (Kennedy, J. F., ed.), vol. 6B, pp. 699–728, VCH Verlagsgesellschaft, Weinheim
32. Hampp, N., Bräuchle, C. & Oesterhelt, D. (1990) Biophys. J. **58**, 83–93
33. Hampp, N., Miller, A., Bräuchle, C. & Oesterhelt, D. (1989) GBF Monogr. ser. **13**, 377–383
34. Hampp, N. & Miller, A. (1990) Werk + Wirken, **1/90**, 39–47
35. Moses, V. (1991) in Biotechnology/The Science and the Business (Moses, V. & Cape, R. E., eds.), pp. 371–378, Harwood Academic Publishers, London
36. Quate, C. F. (1991) Nature (London) **352**, 571
37. Neidleman, S. L. (1989) ASM News **55**, 67–70
38. Zaccai, G. & Eisenberg, H. (1990) Trends Biochem. Sci. **15**, 333–337
39. Oren, A. (1986) Can. J. Microbiol. **32**, 4–9
40. Keradjopoulos, D. & Holldorf, A. W. (1977) FEMS Microbiol. Lett. **1**, 179–182
41. Oren, A. (1983) Curr. Microbiol. **8**, 225–230
42. Klibanov, A. M. (1989) Trends Biochem. Sci. **14**, 141–144
43. Deetz, J. S. & Rozzell, J. D. (1988) Trends Biotechnol. **6**, 15–19
44. Hough, D. W. & Danson, M. J. (1989) Lett. Appl. Microbiol. **9**, 33–39
45. Oxender, D. L. & Graddis, T. J. (1991) in Biotechnology/The Science and the Business (Moses, V. & Cape, R. E., eds.), pp. 153–166, Hardwood Academic Publishers, London.
46. Evans, R. & Kirkland, D. W. (1988) in Evaporites and Hydrocarbons (Schreiber, B. C., ed.), pp. 256–299, Columbia University Press, New York
47. Barbe, A., Grimalt, J. O., Pueyo, J. J. & Albaiges, J. (1990) Org. Geochem. **16**, 815–828
48. Ourisson, G., Rohmer, M. & Poralla, K. (1987) Microbiol. Sci. **4**, 52–57
49. Zajic, J. E. & Akit, J. (1983) in Microbial Enhanced Oil Recovery (Zajic, J. E., Cooper, D. G., Jack, T. R. & Kosaric, N., eds.), pp. 50–54, Pennwell Publishing Co., Tulsa, OK
50. Post, F. J. & Al-Harjan, F. A. (1988) Syst. Appl. Microbiol. **11**, 97–101
51. Sutherland, I. W. (1986) Microbial Sci. **3**, 5–8
52. Gonzalez, M. E., Alarcón, B. & Carrasco, L. (1987) Antimicrob. Agents. Chemother. **31**, 1388–1393
53. Watanabe, Y., Yamamoto, K., Misaki, A. & Hayashida, S. (1987) Agric. Biol. Chem. **51**, 931–932
54. Arad, S(M). (1987) Int. Ind. Biotechnol. (June/July) pp. 281–284
55. Sowers, K. R. & Gunsalus, R. P. (1988) J. Bacteriol. **170**, 998–1002
56. Antón, J., Meseguer, I. & Rodriguez-Valera, F. (1988) Appl. Environ. Microbiol. **54**, 2381–2386
57. Rodriguez-Valera, F., García Lillo, J. A., Antón, J. & Meseguer, I. (1991) in General and Applied Aspects of Halophilic Microorganisms (Rodriguez-Valera, F., ed.), pp. 373–380, Plenum Press, New York
58. Pfiffner, S. M., McInerney, M. J., Jenneman, G. E. & Knapp, R. M. (1986) Appl. Environ. Microbiol. **51**, 1224–1229
59. Clark, J. B., Munnecke, D. M. & Jenneman, G. E. (1981) Dev. Ind. Microbiol. **22**, 695–701
60. Anderson, A. J. & Dawes, E. A. (1990) Microbiol. Rev. **54**, 450–472
61. Lundgren, D. J. & Pfister, R. M. (1964) J. Gen. Microbiol. **34**, 441–446
62. Dawes, E. A. & Senior, P. J. (1973) Adv. Microbiol. Physiol. **10**, 135–266

63. Wallen, L. L. & Rohwedder, W. K. (1974) Environ. Sci. Technol. **8**, 576–579
64. Brandl, H., Gross, R. A., Lenz, R. W. & Fuller, R. C. (1990) Adv. Biochem. Eng. Biotechnol. **41**, pp. 77–93
65. Steinbüchel, A. & Schlegel, H. G. (1991) Mol. Microbiol. **5**, 535–542
66. Steinbüchel, A. & Schubert, P. (1989) Arch. Microbiol. **153**, 101–104
67. Huisman, G. W., Wonink, E., Meima, R., Kazemier, B., Terpstra, P. & Witholt, B. (1991) J. Biol. Chem. **266**, 2191–2198
68. Pries, A., Steinbüchel, A. & Schlegel, H. G. (1990) Appl. Microbiol. Biotechnol. **33**, 410–417
69. Pool, R. (1989) Science **245**, 1187–1189
70. Holmes, P. A. (1985) Phys. Technol. **16**, 32–36
71. Byrom, D. (1987) Trends Biochem. Sci. **5**, 246–250
72. Doi, Y. & Abe, C. (1990) Macromolecules **23**, 3705–3707
73. Fraser, H. M., Sandow, J., Seidel, H. R. & Lunn, S. F. (1989) Acta Endocrinol. (Copenhagen) **121**, 841–848
74. Byrom, D. (1990) in Novel Biodegradable Microbial Polymers (Dawes, E. A., ed.), pp. 113–117, Kluwer Academic Publishers, Dordrecht
75. Pearce, H. (1990) Scientific European **171**, 14–17
76. García Lillo, J. & Rodriguez-Valera, F. (1990) Appl. Environ. Microbiol. **56**, 2517–2521
77. Keeler, R. (1991) R&D Magazine **33**, 52–57
78. Kushner, D. J. (1966) Biotechnol. Bioeng. **8**, 237–245
79. Cooney, C. L. & Wise, D. W. (1975) Biotechnol. Bioeng. **17**, 1119–1124
80. Weimer, P. J. (1986) in General, Molecular and Applied Microbiology (Brock, T. D., ed.), pp. 217–255, John Wiley & Sons, New York

Biochem. Soc. Symp. **58**, 149–169
Printed in Great Britain

Enzymes from thermophilic archaebacteria: current and future applications in biotechnology

Don A. Cowan

Department of Biochemistry and Molecular Biology, University College London, Gower St, London WC1E 6BT, U.K.

Summary

The one guaranteed property of enzymes isolated from extremely thermophilic micro-organisms is their thermostability. Most significantly, almost any such enzyme will be more thermostable than the functionally similar enzyme from a lower temperature source.

Thermostability is not an isolated property: resistance to heat denaturation imparts stability to a number of other denaturing influences (detergents, organic solvents, etc). These characteristics of hyperthermophilic enzymes are the most likely basis for the development of new biotechnological applications.

A limited number of hyperthermophilic enzymes have found application in specialist biotechnological applications; others have visible potential in growing areas of biotechnology. Existing and potential applications are discussed using DNA manipulation enzymes, dehydrogenases, and esterases as examples.

Introduction

Much recent attention has been focused on proteins derived from the extremely thermophilic archaebacteria. Not only do these macromolecules provide a means of probing some of the fundamental properties of protein structure, function and stability, but there is a general assumption that their inherent stability must inevitably lead to valuable contributions to the field of biotechnology. The object of this paper is not to survey the current biotechnological applications of enzymes and to identify where thermophilic enzymes should be applied, but to discuss the way in which the unique properties of these enzymes may be utilized in the development of new biocatalytic processes. The examples given in this paper are designed to be representative, and no attempt is made to provide a comprehensive review of industrial biocatalysis.

A discussion of the biotechnology of a specific group of enzymes or organisms (in this instance, the enzymes derived from the thermophilic representatives of the archaebacteria) can be very conveniently divided into three areas. It is interesting to first consider the specific properties of these enzymes which are (or may be) particularly suited to certain general or specific biotechnological applications. This may give us a clue to the potential development of applied systems. Having considered some of the fundamental properties, a review of the current and historical usage and applications of thermophilic archaebacterial enzymes provides an insight into appreciation by industry of these properties and their willingness to invest in this new technology. Finally, based on a knowledge of the potentially useful characteristics and the current trends in the development of biotechnology, it may be possible to predict either specific applications or more general areas of enzymology which have some likelihood of developing into commercial/industrial projects within the foreseeable future.

In this discussion, and purely for the sake of simplicity and brevity, references to the thermophilic, extremely thermophilic and hyperthermophilic representatives of the archaebacteria are all encompassed under the term 'thermophilic'. While proteins derived from the thermophilic *Thermoplasma* [optimal growth temperature (T_{opt}) = 55 °C] are generally very much less stable than those from the ultra-thermophilic *Pyrodictium* (T_{opt} = 105 °C), there are no fundamental differences in structure or function between either end of the scale.

Properties of enzymes from thermophilic archaebacteria

By 'ecological' necessity, all enzymes derived from the representatives of the thermophilic archaebacteria are thermostable to a greater or lesser extent. It is a fair approximation that any enzyme will demonstrate a reasonable level of thermostability, for example a half-life in the order of hours, at the growth optimum of the organism from which it is derived. This generalization must be tempered by a realization that some enzymes will be much more thermostable (the extracellular hydrolases tend to demonstrate high levels of thermostability at temperatures of 10–20 °C above the growth optimum) while others may appear, at least *in vitro*, to be less thermostable. It should be remembered that for metabolic enzymes, the dual requirements influencing the evolution of protein stability will have been a turnover rate sufficient to provide metabolic control plus sufficient molecular stability to enable a desired metabolic rate to be achieved.

Furthermore, we can now say with a reasonable degree of confidence that any enzyme isolated from an extreme thermophile will almost certainly be more thermostable than a homologous enzyme isolated from a less thermophilic source. This trend can be clearly demonstrated in a comparison of β-galactosidases from various sources (Table 1) and a large number of comparisons of this type with different enzymes are now available in the literature.

The biotechnological consequence of this observation is that where a high level

Table 1. Thermostabilities of mesophilic, thermophilic and hyperthermophilic proteolytic enzymes. While assay conditions will vary considerably, it is assumed that thermostability determinations are carried out under 'near-optimal' conditions of pH, ionic strength, etc.

Source	Growth temperature (°C)	Incubation temperature (°C)	Half-life (min)	Reference
Bacillus subtilis	37	60	10	[1]
Streptomyces rectus	50	82	30	[2]
B. thermoproteolyticus	55	80	60	[3]
B. caldolyticus	72	80	> 480	[4]
Thermus aquaticus	75	80	1800	[5]
Sulfolobus acidocaldarius	70	80	> 2880	[6]
Desulfurococcus mucosus	88	95	75	[7]
Pyrococcus furiosus	94	95	> 1200	[8]

of thermostability is a prerequisite (all other characteristics being equal), the most likely source of an enzyme possessing the desired qualities is a hyperthermophilic archaebacterium. This raises the question of why it has been seen necessary to spend both considerable time and money on the site-directed mutagenesis of other enzymes in order to enhance their molecular stability. The answer is that stability is often of secondary importance and the presence of other highly desirable functional characteristics (activity, specificity, resistance to inhibition, etc), which are not necessarily present in a thermophilic homologue, justifies this approach.

Extreme thermostability *per se* is not necessarily a valuable characteristic in the industrial application of enzymes. However, some of the consequences of protein thermostability are possibly of more relevance in the future biotechnological potential of thermophilic enzymes.

One of the (inevitable?) consequences of protein thermostability appears to be a high degree of resistance to other potential denaturants: detergents, chaotropic agents such as urea and guanidinium hydrochloride, organic solvents, oxidizing agents, and possibly extremes of pH. There is evidence that thermostability also imparts resistance to enzymic cleavage (proteolysis) and to chemical reactions causing irreversible protein degradation at high temperatures [9].

The resistance of thermostable proteins to denaturation by organic solvents appears to be a general characteristic. It has been demonstrated [10] that a clear positive correlation exists between thermostability and resistance to denaturation in biphasic organic/aqueous solvent systems, both for populations of proteins and for single purified enzymes (Figs. 1*a* and *b*). The organic solvent system chosen for this study, 50:50 (v/v) butanol:buffer, is a significant protein denaturant. The enzyme, localized in the aqueous phase, is exposed both to dissolved organic phase and to liquid–liquid interfaces. Each data point on this graph represents a cellular protein extract (representing a population of proteins). The index of thermostability is based on the degree of protein precipitation after heating for a defined period at a defined temperature. The index of organic solvent stability is based on similar criteria.

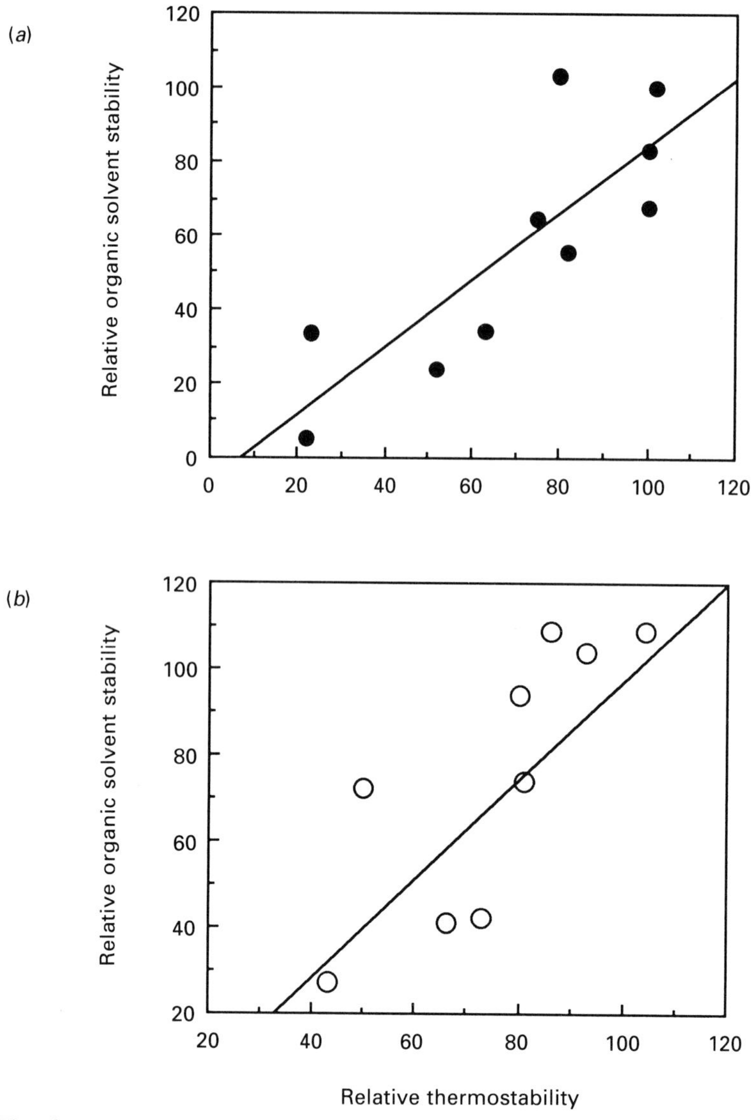

Fig. 1. A correlation between protein thermostability and resistance to denaturation in organic solvents. Results obtained using (*a*) protein populations derived from whole cell extracts and (*b*) purified proteolytic enzymes.

As indicated by a number of other workers (e.g. [11]), this general correlation also holds for single purified enzymes (Fig. 1*b*). The data were obtained from a range of purified proteolytic enzymes derived from mesophiles, thermophiles and ultrathermophiles, compared under the same conditions as for Fig. 1*a*. The index of stability is based on loss of enzymic activity resulting from exposure to temperature or the organic solvent system.

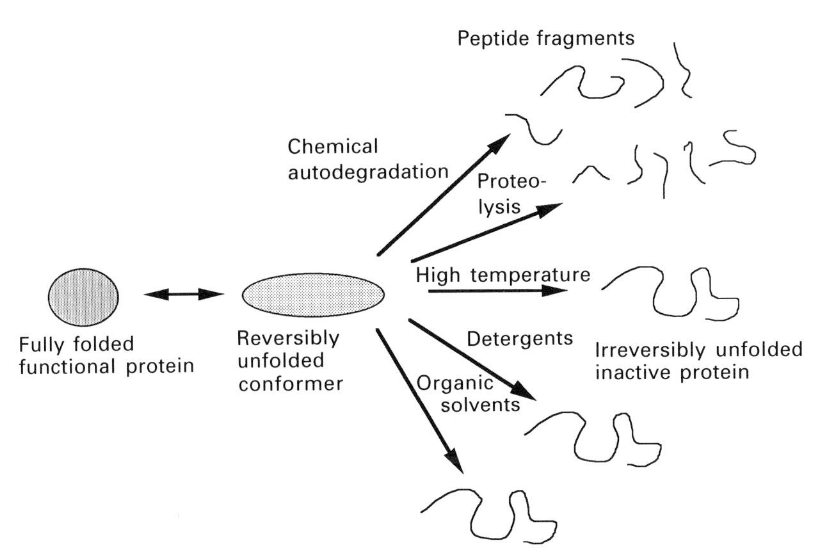

Fig. 2. A schematic view of the pathways to irreversible enzyme inactivation.

While the detailed molecular mechanisms of the denaturation of proteins in organic solvents, detergents, etc are poorly understood, the evidence that thermostable proteins show enhanced resistance to all these agents implies that some aspect of the unfolding pathway is common to all. It seems likely that the first step of protein unfolding is the rapid reversible conformational transition which is a reflection of the normal flexibility of the protein. Loss of tertiary structure via denaturation (by temperature, detergent, solvent, etc) or proteolysis will only progress from the unfolded conformer (Fig. 2). Restriction of the initial reversible conformational transitions, a consequence of thermostabilization, will proportionally reduce the tendency of the protein to partake in future irreversible unfolding steps. The resistance of thermostable proteins to proteolysis supports this scheme, since it is well known that only unfolded proteins are susceptible to proteolytic attack. The observation that chemical degradation of proteins, including the deamination of asparagine residues and peptide cleavage at Asn-Xaa bonds, proceeds more rapidly in pre-denatured proteins (Hensel, pp. 127–133 in this volume), also supports this mechanism.

While molecular stability under a wide range of different conditions is generally assumed to be an advantageous property, another apparent consequence of thermostability is, at least for the biotechnologist, a disadvantage. A common misconception is that a consequence of thermophilicity is the ability to attain very rapid reaction rates by carrying out reactions at high temperatures. This misconception is based on a simple Arrhenius law extrapolation from the properties of mesophilic enzymes, presuming that if the reaction rate of a mesophilic enzyme is X at 37 °C, then the equivalent hyperthermophilic enzyme should be 32X at around 90 °C. We now know this to be incorrect in almost all instances and, for sound physical reasons, thermophilic enzymes at their 'optimum' temperatures exhibit turnover rates of the same order of magnitude as do mesophilic enzymes at their

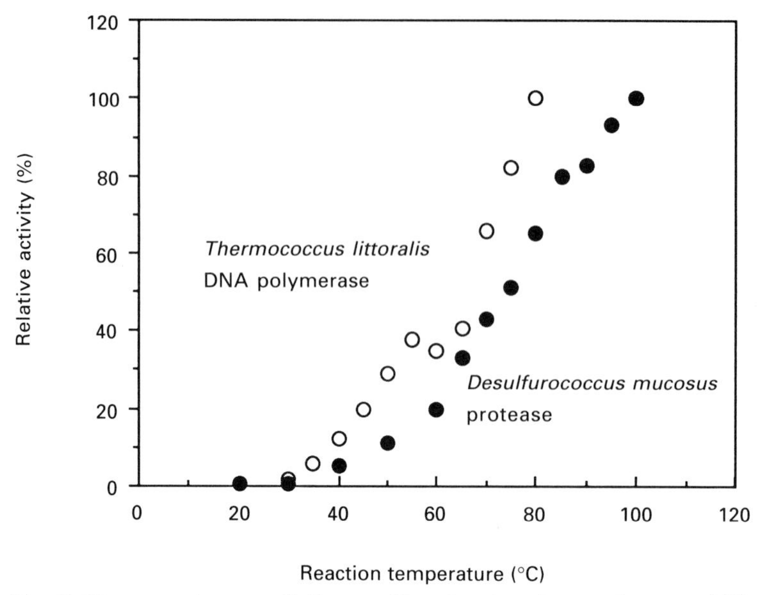

Fig. 3. Temperature–activity profiles for two hyperthermophilic enzymes. Data are derived from New England Biolabs technical literature (*Thermococcus littoralis* DNA polymerase) and [7] (*Desulfurococcus* protease).

'optimum' temperatures. The mechanism of this apparent limitation of activity is, at least in part, the restraint in conformational flexibility which is thought to be coincident with enhanced thermostability [12]. Evidence supporting this relationship is still sketchy and there is great potential for detailed structure/function/stability studies. It should not be assumed, for example, that the conformational restraints will be consistent in different parts of the potential tertiary structure.

In practice, when using extremely thermophilic enzymes, it is desirable to operate reactions as near to their 'optimum' temperatures as possible in order to attain reasonable rates of catalysis. This is clearly shown in the temperature–activity profiles (Fig. 3) of two hyperthermophilic enzymes. With so-called 'temperature optima' at around 90 °C and 100 °C respectively, lowering the reaction temperature by 20 °C causes respective 36% and 65% reductions in reaction rates for *Desulfurococcus mucosus* protease and *Thermococcus littoralis* DNA polymerase. Because of operational problems such as reactant or product instability, it may not always be either feasible or desirable to carry out biocatalytic reactions at such elevated temperatures, producing the inevitable loss of catalytic efficiency if reactions are carried out at lower and more acceptable temperatures. On the positive side, the catalytic rates are so slow at room temperature (estimated to be $< 2\%$ of 'optimum activity') that, for most purposes, they may be ignored.

There is one area where significant enhancement of reaction rates can be achieved by using thermophilic enzymes at high temperatures. These are the cases where the substrate itself is rendered more susceptible to catalysis at higher temperatures. The best examples are the hydrolytic degradation of polymeric

Table 2. Cloning of protein genes from thermophilic archaebacteria. Abbreviation used: EF, elongation factor.

Enzyme/Protein	Source organism	Expression	Reference
Aspartate aminotransferase	*Sulfolobus solfataricus*	No	[13]
ATPase subunits	*Sulfolobus acidocaldarius*	No	[13]
Citrate synthase	*Thermoplasma acidophilum*	*E. coli*	[14]
DNA polymerase	*Thermococcus littoralis*	Yes	a
DNA polymerase	*Pyrococcus furiosus*	Yes?	b
EF-2 protein	*S. acidocaldarius*	No	[15]
β-Galactosidase	*S. solfataricus*	*E. coli*	[16]
Glutamine synthetase	*S. solfataricus*	No	[17]
Glyceraldehyde-3-phosphate dehydrogenase	*Pyrococcus woesei*	*E. coli*	[18]
Protease (Thermopsin)	*S. acidocaldarius*	No	[6]
RNA polymerase	*S. acidocaldarius*	No	[19]

[a] Cloned by New England Biolabs, MA, U.S.A. (N.E.B. Technical Literature).
[b] Cloned by Stratagene Ltd, La Jolla, CA, U.S.A. (Stratagene Technical Literature).

substrates, particularly proteolytic degradation of protein and amylolytic degradation of starch. Industry has taken full advantage of the latter, particularly in the high temperature saccharification processes currently employed in most enzymic starch hydrolysis plants.

Another problem inherent in the potential use of enzymes from thermophilic archaebacteria should also be highlighted. The biomass yields from archaebacterial fermentation are extremely low (typically 0.1–1.0 g/l wet weight of archaebacterial biomass compared with 5–30 g of biomass/l for common eubacteria). This is a major impediment to any direct application of thermophilic archaebacteria, whether as whole cell biocatalysts, sources of enzymes, or sources of other biomolecules. It is therefore a basic assumption of this discussion of the biotechnology of archaebacterial enzymes that, in most (possibly all) examples, the successful application of any archaebacterial enzyme will follow the cloning and over-expression of the protein. On an optimistic note, this impediment to use is shrinking rapidly. The cloning and over-expression of hyperthermophilic proteins is now known to be a perfectly feasible objective, as demonstrated by the growing collection of examples reported in the literature (Table 2).

Commercial applications of archaebacterial enzymes

In surveying the commercially successful biotechnological applications of enzymes from thermophilic archaebacteria, it is possible to summarize these in a single application: the polymerase chain reaction (PCR). The central core of this well publicized and widely used technique is the application of thermostable DNA

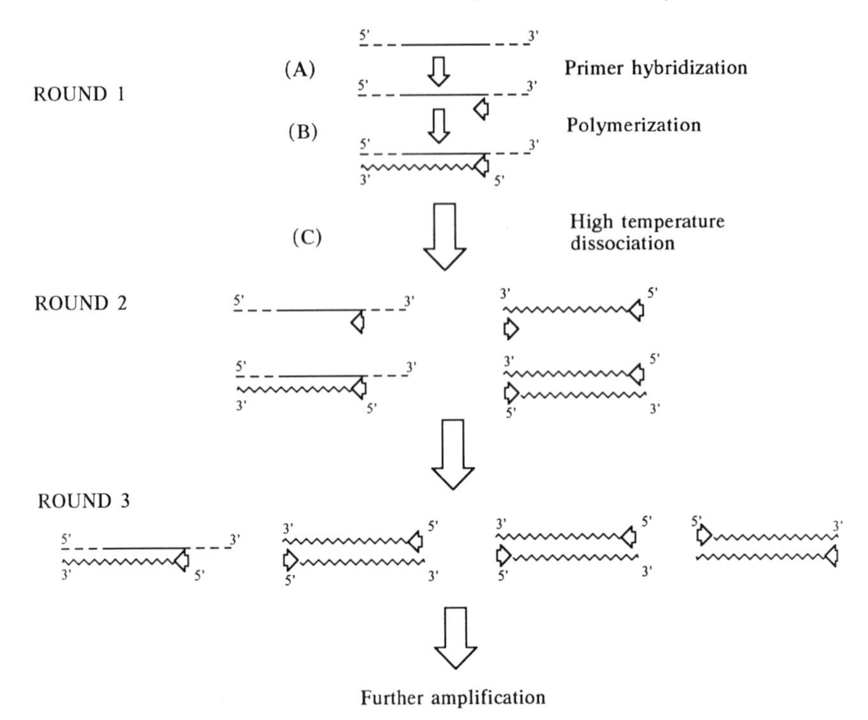

Fig. 4. Amplification of specific DNA sequences using the polymerase chain reaction.

polymerases, providing a rapid and efficient exponential amplification of specific sequences of DNA. As shown in Fig. 4, the protocol involves the hybridization of specific primers (A), the DNA polymerase-catalysed extension of nucleic acid sequences complementary to the template (B), and the high temperature dissociation of the nascent oligonucleotide chain from the template (C), thus allowing both nascent and template strands to act as templates for the next cycle of amplification.

The technique as originally developed with the Klenow fragment of *Escherichia coli* DNA polymerase required repetitive additions of this thermolabile enzyme, since all activity was lost during the high temperature dissociation step (typically 90–95 °C). The automation of PCR by incorporation of the thermostable DNA polymerase from the thermophilic eubacterium *Thermus aquaticus* (where patent protection on both the method and its applications have been assigned to the Cetus Corporation of California [20]) was a major development in the technology.

The thermostability of *Thermus aquaticus* (*Taq*) polymerase is marginally sufficient under operational conditions and a further recent development is the appearance of extremely thermostable DNA polymerases from two different hyperthermophilic archaebacteria (*T. littoralis* and *P. furiosus*). The comparative thermostabilities of these enzymes are shown in Fig. 5. The high profile and commercial success of thermostable DNA polymerases has stimulated a plethora of studies and other thermophilic sources of this enzyme including *Thermotoga maritima* [21] (eubacterial hyperthermophile), *Thermoplasma acidophilum* [22], *Methanobacterium thermoautotrophicum* [23], *Sulfolobus acidocaldarius* [24] and *Sulfolobus solfataricus* [25]

Table 3. Properties of the 'commercial' DNA polymerases.

Enzyme/ Source	Yield[a]	Thermo- stability	Processivity	3′–5′ Exo- nuclease proof- reading	Fidelity (mutations per base duplication)	Reference
Klenow (*Escherichia coli*)	Moderate	Low	Good	Yes	8×10^{-5}	[26]
Bst (Bacillus)	Good	Mod	Very good	?	?	[27]
Taq (Thermus)	Moderate	High	Good	No	2×10^{-4}	[26,28]
Tth (Thermus)	Moderate	High	?	?	?	—
Tub (Thermus)	Moderate	High	?	?	?	—
Thermococcus	Very low	Very high	Very good	Yes	?	[b]
Pyrococcus	Very low	Very high	Very good	Yes	$< 1.7 \times 10^{-5}$	[b]

[a] Yield values are for the parent (non-recombinant) strains.
[b] Data obtained from New England Biolab and Stratagene technical literature.

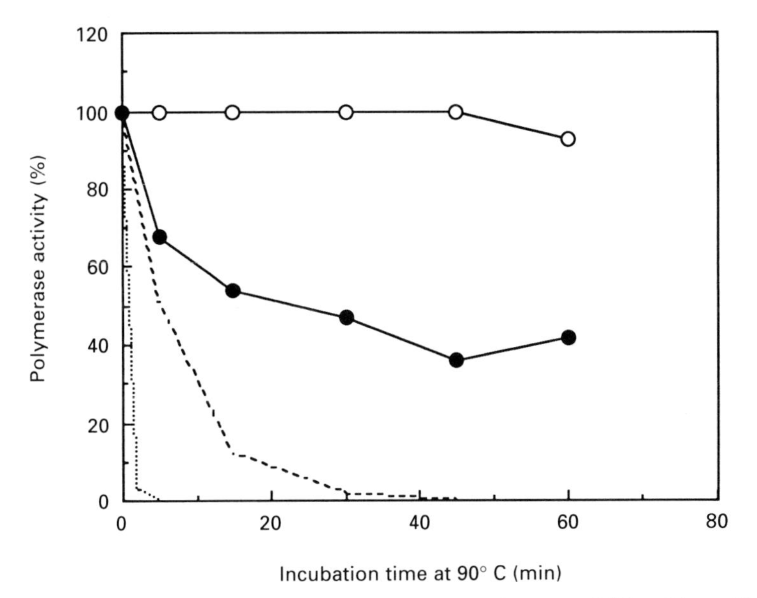

Fig. 5. Comparative thermostabilities of mesophilic *(E. coli)*, thermophilic *(T. aquaticus)* and hyperthermophilic *[Thermococcus* (●) and *Pyrococcus* (○)*]* DNA polymerases at 90 °C. Data are derived from Stratagene technical literature (*Pyrococcus* and *Thermococcus* DNA polymerases) and [77] for *Taq* polymerase.

have been investigated in some detail, but other DNA polymerases are not commercially available. However, a high degree of thermostability is not the only necessity for successful application of DNA polymerases in PCR. Other factors such as processivity and proofreading (3′–5′ exonuclease) activity are also important (Table 3).

For the DNA polymerases from the hyperthermophilic archaebacteria, it is quite justified to assume that commercial feasibility is totally dependent on prior success in the cloning and over-expression of the enzymes. The manufacturers of the *Pyrococcus* and *Thermococcus* enzymes have been successful in this objective, despite such surprising hurdles as the presence of introns in the protein coding sequences [29].

Future applications of enzymes from thermophilic archaebacteria

The field of archaebacterial enzymology is in the early stages of development. At this point, the number of enzymes from thermophilic archaebacteria which have been studied in any detail (Table 4) barely exceeds the total number of named species. While it is therefore too early to use the distribution of enzymes as an indication of their potential, there is every reason for assuming that these organisms will contain as wide a range of different activities as any other diverse group of micro-organisms. Furthermore, certain activities and enzyme systems will be totally unique to this group of organisms, an excellent example being the (as yet uncharacterized) enzymes of the biphytanyl tetraether lipid biosynthetic pathways.

The success of thermophilic enzymes in PCR provides a valuable insight on one of the more likely paths of future development in biotechnology. PCR is an excellent example of a 'niche' application, where the specific properties of the enzyme are complementary to the highly specific requirements of a process. Based on this example, it is predicted that successful applications are more likely to arise where the specific and unique characteristics of the enzymes can be used to advantage.

In considering possible commercial and industrial applications of thermophilic enzymes, one inevitably turns to the current biotechnological enzyme applications. Over 75% of the world usage of enzyme (currently estimated at over £800 million) is to be found in a very limited number of industries (Fig. 6). The detergent industry, using alkalophilic proteases; the starch industry, using amylases, amyloglucosidases and glucose isomerases, and the dairy industry using milk-clotting enzymes contribute the majority of this portion.

In a closer analysis of the status of the most important of these industrial enzymes (Table 5), it is logical to put the hypothetical question: Could a thermostable alternative replace the enzyme in current use? The answer to this question is, in many instances, yes. However, a more important question is: Is it commercially feasible to replace the enzyme in current use with a thermostable alternative? The answer here is a qualified no. These conclusions are based on several observations: (i) certain industrial biocatalytic processes are incompatible with high-temperature operation (e.g. cheese-making); (ii) commercial trends favour low-temperature operation (e.g. domestic detergents), (iii) enzymes of sufficient thermostability are already in use (e.g. saccharification of starch); (iv) there is no obvious process advantage in increasing the reaction temperature (e.g. biosynthetic production of fine chemicals such as amino acids and penicillins) and (v) many of the large industrial biocatalyst users are committed to existing low-temperature operation through investment in specific plant and equipment, and the advantage of higher temperature operation would be insufficient to offset the costs of replacing that equipment. In summary, the

Table 4. Detailed studies on enzymes from thermophilic archaebacteria.

Organism	Enzyme	Reference
Acidianus brierleyi	Sulphur oxygenase	[57]
Archaeglobus fulgidus	ATP sulphurylase	[36]
Desulfurococcus mucosus	Protease	[7]
	RNA polymerase	[56]
	Transglucosylase	[45]
Pyrococcus furiosus	α-Amylase	[33]
	α-Glucosidase	[44]
	Hydrogenase	[49]
	Protease	[8]
	Pullulanase	[55]
Pyrococcus woesei	Glyceraldehyde-3-phosphate dehydrogenase	[18]
Sulfolobus acidocaldarius	ATPase	[34]
	Citrate synthase	[39,40]
	Carboxylesterase	[41]
	DNA polymerase	[24]
	Glyceraldehyde-3-phosphate dehydrogenase	[18]
	Isocitrate dehydrogenase	[50]
	Malate dehydrogenase	[51]
	NADH dehydrogenase	[53]
	Polyphosphate kinase	[54]
	Protease	[6]
	RNA polymerase	[56]
	Topoisomerase I	[58]
Sulfolobus solfataricus	S-Adenosylmethionine synthetase	[30]
	Alcohol dehydrogenase	[31]
	Aminopeptidase	[32]
	ATPase	[35]
	Aspartate aminotransferase	[37]
	Citrate synthase	[38]
	DNA polymerase	[25]
	Fumarase	[43]
	β-Galactosidase	[46]
	Glucose dehydrogenase	[48]
	Malic enzyme	[52]
Thermoplasma acidophilum	Dihydrolipoamide dehydrogenase	[42]
	DNA polymerase	[22]
	Glucose dehydrogenase	[47]
	Malate dehydrogenase	[51]
	RNA polymerase	[56]
	Succinate thiokinase	[40]
Thermococcus celer	α-Glucosidase	[44,45]
	β-Glucosidase	[45]
	Protease	[21]
Thermoproteus tenax	Glyceraldehyde-3-phosphate dehydrogenase	[18]
	RNA polymerase	[56]

Table 5. Current usage of thermostable enzymes in major commercial and industrial processes.

Industry/ Application	Enzyme used	Thermo-stable enzyme in use	More thermostable enzyme required
Starch hydrolysis	α-Amylase	Yes	No
	Amyloglucosidase	No	(Yes)
	Glucose isomerase	No	(Yes)
	Pullulanase	No	No
Bread making	β-Amylase	No	No
Protein hydrolysis			
Detergents	Alkaline protease	No	No
Cheese-making	Rennin, *Mucor* rennin	No	No
Tanning	Alkaline protease	No	No
Clarification of wine and beer	Pectinase	No	No
Amino acid production	Various	No	No
Antibiotic semi-synthesis	Penicillin V acylase	No	No
DNA cloning	Restriction endonucleases	Some	(Yes)

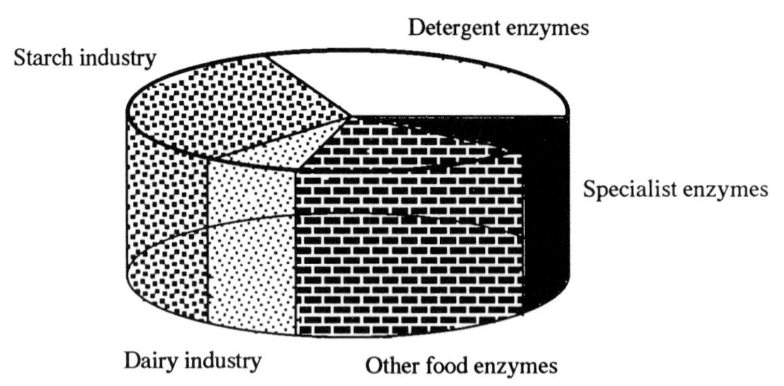

Fig. 6. Major industrial users of enzymes in the 1980s.

conditions under which a thermophilic enzyme would be a serious contender as a replacement for any 'bulk' commercial enzyme might include many (if not all) of the following: (i) Availability in similar (or equivalent) quantity with an equally reliable supply; (ii) A similar or lower price (per unit of activity); (iii) Significantly better performance in many respects (not just stability) and (iv) No major alterations to existing plant and equipment (or investment in new plant) required.

This assessment leads us to the conclusion that the future for thermophilic

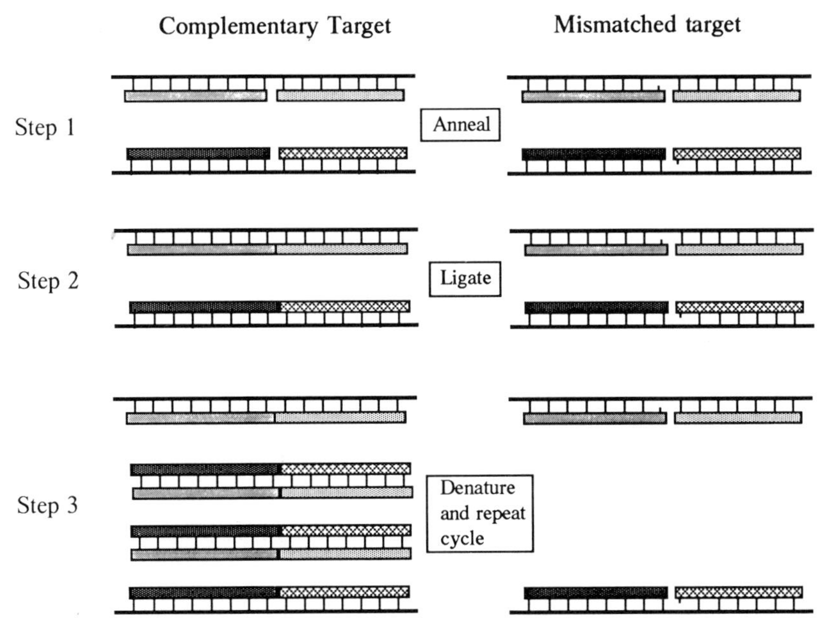

Fig. 7. DNA amplification using the ligase chain reaction as a method for detecting genetic lesions.

archaebacterial enzymes will not be as replacements for existing enzymes, particularly in the high-volume, low-cost enzyme market. While this may exclude a large portion of the industrial market, the residual fraction comprises a multitude of more minor applications and other major applications loom in the near future. In some of these 'niche' applications, thermophilic enzymes may yet play a major role, particularly in those instances where the unique properties that these enzymes possess offer significant process advantages.

It is not possible to provide a detailed assessment of each current industrial/analytical enzyme application in an attempt to quantify the potential 'penetration' of thermophilic enzymes. Rather, a few specific examples are provided here as a demonstration of the way in which unique properties of thermophilic enzymes, particularly the property of structural stability, can be applied to certain areas of biotechnology.

Example 1—the ligase chain reaction

The first example is the recent development [59] of a ligase-dependent DNA amplification technique (the ligase chain reaction). In the ligase chain reaction (Fig. 7), the DNA template is heat denatured and four complementary oligonucleotide primers are hybridized to the target at near the melting temperature (t_m). Thermostable DNA ligase is then added and ligation between adjacent primers proceeds; the success of the technique relies on the fact that at the high temperature, only adjacent and perfectly complementary oligonucleotides will be ligated together. The ligation products are heat denatured and form new target templates for the next cycle, resulting in an exponential amplification. Oligonucleotides with a single base

Table 6. Substrate specificities of sec-alcohol dehydrogenases from thermophilic archaebacteria.

Organism	Growth temperature (°C)	Cofactor	Relative activity on: Ethanol	2-Butanol	Cyclo-hexanol	Benzyl alcohol
Sulfolobus solfataricus[a]	87	NAD	100	188	410[b]	89
Hyperthermus butylicus	96	NADP	100	76	174	0

[a] Data from [31]
[b] Value for 3-methylcyclohexanol
[c] Data from L. Galabe & D. A. Cowan, unpublished work.

mismatch will not ligate and will amplify only in a linear fashion. The potential use of this technique for the identification of gene defects is considerable and has been demonstrated in the detection of β-globin mutants [59,60].

The requirements for a thermophilic and thermostable DNA ligase are also obvious—the ligation must be carried out at near the DNA melting temperature (a minimum of 65 °C) and the enzyme must be stable during the subsequent dissociation step (carried out at about 92 °C). To date, the only thermostable DNA ligase to be reported is derived from the thermophilic eubacterium *Thermus thermophilus* [61]. This enzyme has been cloned and is commercially available.

Example 2—alcohol dehydrogenases

The secondary alcohol-specific alcohol dehydrogenases are a group of enzymes which undoubtedly have an important future in industrial biocatalysis. These enzymes are capable of catalysing either the oxidation of alcohols or the reverse reaction, the reduction of aldehydes and ketones. There is considerable interest in the use of such enzymes in the chemical synthesis industries, particularly the pharmaceutical industries, where the production of chiral synthons is an increasingly important step in the synthesis of chirally pure pharmaceutical agents [62–65]. While alcohol dehydrogenases are widely distributed and well characterized, only two examples of thermostable alcohol dehydrogenases derived from hyperthermophilic archaebacteria are currently known (Table 6). The *Sulfolobus* enzyme is NAD-specific [31] whereas the *Hyperthermus* alcohol dehydrogenase has a dual co-factor specificity with a marginal preference for NAD (L. Galabe & D. A. Cowan, unpublished work). The capacity to reduce aromatic aldehydes and ketones to their respective chiral alcohols, an activity possessed by the *Sulfolobus* alcohol dehydrogenase, is particularly valuable.

In the reactions carried out by alcohol dehydrogenases, the presence of significant concentrations of miscible organic solvents can be a significant problem [66]. For instance, both the substrate and product may be hydrophilic organic molecules (alcohols, aldehydes and ketones), and such organic 'solvents' are often potent denaturants [67]. Where reaction systems are configured to optimize reaction

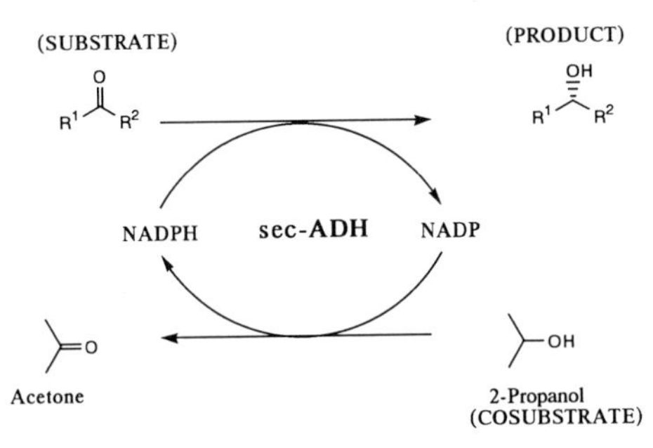

Fig. 8. A co-factor recycle system for secondary-alcohol dehydrogenase (sec-ADH)-catalysed production of chiral alcohols.

Table 7. Examples of thermophilic secondary-alcohol dehydrogenase catalysed chiral reductions.

Substrate	Product	% e.e.	Absolute configuration	Reference
		48	*R*	[69]
		96	*S*	[69]
		98	*S*	[69]
		99	*S*	a
		69	?	a
	and			[70]

[a] From Cowan & Plant (1991) Abstr. 201st ACS meeting, Atlanta, April 1991, **201**, 99.

productivity, both substrate and product may be present at relatively high concentrations. Furthermore, the high cost of NADH and NADPH ensures that, for all but laboratory-scale operations, a system for recycling the oxidized co-factor is essential. One effective recycle option is to include an alcohol co-substrate which, in being oxidized by the enzyme, regenerates the reduced co-factor (Fig. 8). However, this results in even higher levels of potentially denaturing organic components in a standard reaction mixture.

The presence of high concentrations of miscible organic solvents necessitates the use of enzymes stable in organic solvents. This approach has been used in the biocatalytic reduction of a variety of pure chiral alcohols (some examples are given in Table 7) using a moderately thermostable alcohol dehydrogenase isolated from a eubacterial thermophile *Thermoanaerobium brockii* [68–70]. However, the stability of the enzyme in this reaction is marginal, where reaction temperatures greater than

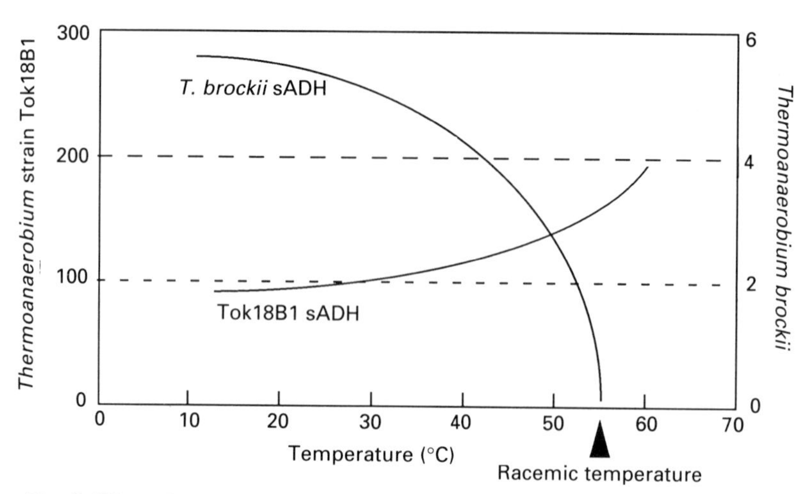

Fig. 9. The effect of reaction temperature on the stereoselectivity of secondary-alcohol dehydrogenase-catalysed oxidation of (*R*) and (*S*)-2-pentanols. Abbreviation used: sADH, secondary-alcohol dehydrogenase. Values quoted are the ratios of the specificity constants for the (*R*) and (*S*) substrates: $k_{cat}/K_m(S)/k_{cat}/K_m(R)$.

35 °C and organic solvent concentrations of more than about 10% induce rapid enzyme inactivation (A. L. Plant & D. A. Cowan, unpublished work). Early reductive biotransformation trials with *Sulfolobus* sp. [71] suggest that the hyperthermophilic enzymes may be excellent alternatives.

There are reports that reaction temperature may influence the stereoselectivity of alcohol dehydrogenase-catalysed reactions [72]. Using a sec-alcohol dehydrogenase derived from a *Thermoanaerobium*-like organism, we have demonstrated that an increase in reaction temperature of 30 °C to 60 °C enhanced the enantioselectivity of the enzyme by a factor of two in the oxidation of (*R*)- and (*S*)-propan-2-ols (Fig. 9). Conversely, in a similar reaction with *Thermoanaerobium brockii* sec-alcohol dehydrogenase (Fig. 9), a reduction in reaction temperature strongly favoured a higher degree of enantioselectivity. Clearly, the optimization of individual bioconversions will require careful selection of enzyme and reaction conditions.

Without doubt, we can expect to see a rapid increase in laboratory studies using thermophilic alcohol dehydrogenases for biotransformation purposes. The appearance of the archaebacterial enzymes in commercial quantities would provide a further stimulus to these developments.

Example 3—esterases

Esterases are widely distributed enzymes and mesophilic examples have already found valuable applications in organic biosynthesis [62,64,73–75]. In aqueous solution, the esterases catalyse the hydrolytic cleavage of esters forming the constituent acid and alcohol. However, in reaction systems where the aqueous component of the solvent is reduced sufficiently, non-hydrolytic reactions can be stimulated (Fig. 10). In relatively low water content (i.e. sufficiently low to reduce the

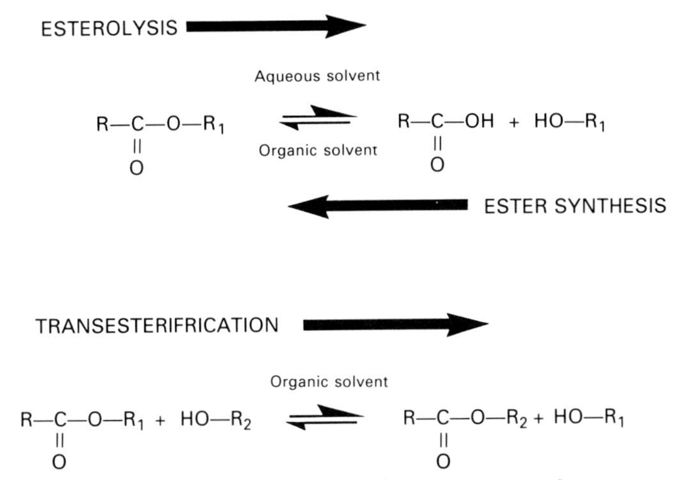

Fig. 10. Esterase-catalysed reactions in aqueous and non-aqueous solvent systems.

thermodynamic water activity) the synthetic reaction can be promoted. Transesterification, also thermodynamically unfavoured in aqueous solution, can be greatly enhanced. Both the reactants and products of transesterification are usually highly soluble in an organic phase, and the reactants may even form the organic phase themselves. Clearly, it is essential that the biocatalyst is stable under the solvent conditions used.

Some excellent examples of the potential of ester-catalysed organic-solvent-based reactions have been reported. The work by Lokotch *et al.* [76] on the resolution of the stereoisomers of D,L-menthol is a clear demonstration of the feasibility of using organic-solvent-based esterase-catalysed reactions. While it is possible to carry out the reaction in aqueous solution [Fig. 11(*a*)], the racemic alcohol must first be converted chemically into the racemic acetate. Stereospecific enzymic hydrolysis of the racemic acetate yielded unreacted D-menthyl acetate and chirally pure L-menthol, which could be separated on the basis of differing hydrophobicities. The optimal activity in aqueous solution was at 50 °C.

However, if the process was performed as a transesterification reaction with either triacetin or acetic acid in 100% isooctane [Fig. 11(*b* & *c*)], the unmodified racemic menthol could be used directly, the primary product being the L-menthyl acetate. On separation, this can be directly hydrolysed to the pure L-isomer. The transesterification reaction occurred much more slowly at 50 °C than the hydrolytic reaction in water at the same temperature. However, in the organic solvent, the esterase was found to be stable at 90 °C where the reaction rates were comparable.

These reactions were carried out using a stable esterase of mesophilic origins. The advantages of using more thermostable esterases lie in the ability to utilize higher reaction temperatures (where substrate/product solubilities are higher and viscosities are lower), higher concentrations of potentially denaturing substrates and a wider range of solvent systems.

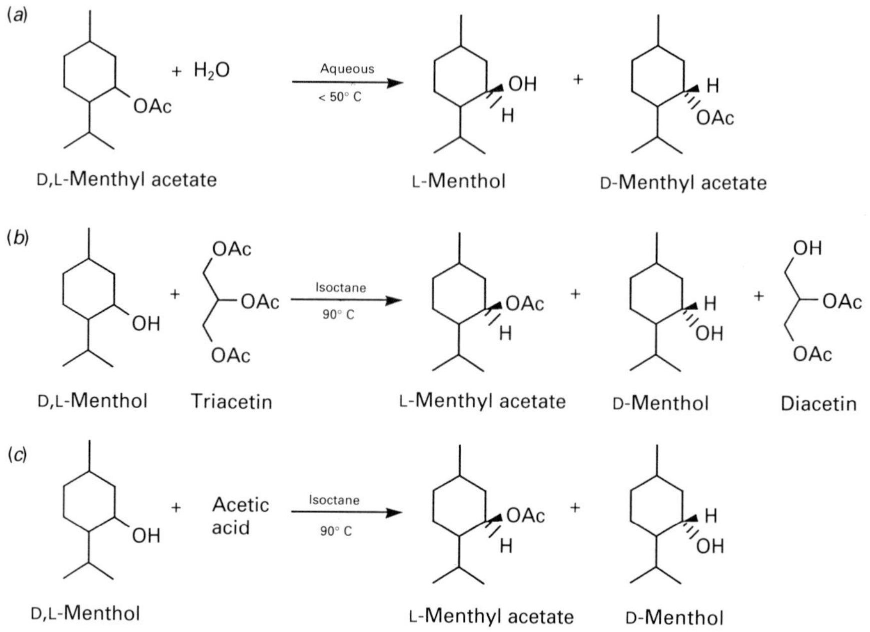

(a)

OAc + H_2O —Aqueous, < 50° C→ OH H + H OAc

D,L-Menthyl acetate L-Menthol D-Menthyl acetate

(b)

OH + OAc OAc OAc —Isoctane, 90° C→ OAc H + H OH + OH OAc OAc

D,L-Menthol Triacetin L-Menthyl acetate D-Menthol Diacetin

(c)

OH + Acetic acid —Isoctane, 90° C→ OAc H + H OH

D,L-Menthol L-Menthyl acetate D-Menthol

Fig. 11. Esterase-catalysed resolution of chiral menthols. Reproduced from [76] with permission.

Conclusions

There will undoubtedly be many other examples of biocatalytic systems where stability and/or function at high temperatures might impart a significant process advantage. For example, enzymes such as aldolases and hydrogenases (in organic semi-synthesis), glucosidases (in oligosaccharide synthesis), amidases and peptidases (in peptide synthesis) and phospholipases (in phospholipid synthesis) are all possible contenders. In particular, reactions where the normal equilibrium is reversed often require potentially destabilizing conditions (high substrate concentrations, addition of solvents, etc), and many may be more efficiently promoted by the use of more stable biocatalysts.

In preparing this document, I hope that I have been successful in combining both optimism and realism. The enthusiasm of the scientist is always tempered by the harsh reality of the marketplace and the transition from discovery to application is at best slow and uncertain. However, we should not allow the difficulties of this transition to inhibit our search for new biocatalysts and for novel approaches to the application of existing ones. Such research, despite being 'application-driven' can also enhance our fundamental knowledge of molecular structure and processes.

The biotechnological prospects of enzymes from thermophilic archaebacteria are far from bleak. While they may not be successful in replacing the high-volume, low-cost enzymes which comprise most of 'industrial enzymology', there are a multitude of smaller more specialized applications where the unique properties of thermophilic enzymes may prove to be advantageous. With the increasing availability of these enzymes, more thought will be applied *a priori* to the design of processes to take advantage of their molecular stability.

References

1. Fullbrooke, P. D. (1983) in Industrial Enzymology (Godfrey, T. & Reichelt, J., eds.), pp. 8–10, Nature Press, New York
2. Mizusawa, K. & Yoshida, F. (1973) J. Biol. Chem. **248**, 4417–4423
3. Endo, S. (1962) Hakko Kogaku Zasshi **40**, 346–353
4. Heinen, U. J. & Heinen, W. (1972) Arch. Mikrobiol. **82**, 1–23
5. Cowan, D. A. & Daniel, R. M. (1982) Biochim. Biophys. Acta **705**, 293–305
6. Lin, X.-L. & Tang, J. (1990) J. Biol. Chem. **265**, 1490–1495
7. Cowan, D. A., Smolenski, K. A., Daniel, R. M. & Morgan, H. W. (1987) Biochem. J. **247**, 121–133
8. Connaris, H., Cowan, D. A. & Sharp, R. (1991) J. Gen. Microbiol. **137**, 1193–1199
9. Ahearn, T. J. & Klibanov, A. M. (1985) Science **228**, 1280–1284
10. Owusu, R. K. & Cowan, D. A. (1989) Enzyme Microb. Technol. **11**, 568–574.
11. Guargliardi, A., Rossi, M. & Bartolucci, S. (1989) Chim. Oggi (May), 31–36
12. Vihinen, M. (1987) Protein Eng. **1**, 477–480
13. Cubellis, M. V., Rozzo, C., Nitti, G., Arnone, M. I., Marino, G. & Sannia, G. (1990) Eur. J. Biochem. **186**, 375–381
14. Sutherland, K. J., Henneke, C. M., Towner, P., Hough, D. W. & Danson, M. J. (1990) Eur. J. Biochem. **194**, 839–844
15. Schröder, J. & Klink, F. (1991) Eur. J. Biochem. **195**, 321–327
16. Little, S., Cartwright, P., Campbell, C., Prenneta, A., McChesney, J., Mountain, A. & Robinson, M. (1989) Nucleic Acids Res. **17**, 7980
17. Sanangelantoni, A. M., Barbarini, D., Di Pasquale, G., Cammarono, P. & Tiboni, O. (1990) Mol. Gen. Genet. **221**, 187–194
18. Zwickl, P., Fabry, S., Bogedain, C., Haas, A. & Hensel, R. (1990) J. Bacteriol. **172**, 4329–4338
19. Pühler, G., Lottspeich, F. & Zillig, W. (1989) Nucleic Acids Res. **17**, 4517–4534
20. Mullis, K. B. (1987) Process for Amplifying Nucleic Acid Sequences. U.S. Pat. Appl. 4,683,202
21. Simpson, H. D., Coolbear, T., Vermue, M. & Daniel, R. M. (1990) Biochem. Cell Biol. **68**, 1292–1296
22. Chinchaladze, D. Z., Prangishivili, D. A., Kachabaeva, L. A. & Zaalishvili, M. M. (1986) Mol. Biol. **19**, 1193–1201
23. Klimczak, L. T., Grummt, F. & Burger, K. J. (1986) Biochemistry **25**, 4850–4855
24. Elie, C., De Recondo, A. M. & Forterre, P. (1989) Eur. J. Biochem. **178**, 619–626
25. Rossi, M., Rella, R., Pensa, M., Bartolucci, S., De Rosa, M., Gambacorta, A., Raia, C. A. & Dell'Aversano, N. (1986) Syst. Appl. Microbiol. **7**, 337–341
26. Keohavong, P. & Thilly, W. G. (1989) Proc. Natl. Acad. Sci. U.S.A. **86**, 9253–9257
27. Puchkina, P. A., Kaboev, O. K., Pogenova, P. G. & Khraptsova, G. I. (1988) U.S.S.R. Pat. Appl. SU 1311251
28. Sambrook, J., Fritsch, E. F. & Maniatis, T. (1989) in Molecular Cloning, A Laboratory Manual, 2nd edn., Chapter 5, pp. 5.50–5.51, Cold Spring Harbour Press, U.S.A.
29. Mathur, E. J. (1991) Abstr. N.Y. Acad. Sci. (ACS meeting, Atlanta, April, 1991) **201**, 100
30. Porcelli, M., Cacciapuoti, G., Carteni-Farina, M. & Gambacorta, A. (1988) Eur. J. Biochem. **177**, 273–280

31. Rella, R., Raia, C. A., Pensa, M., Pisani, F. M., Gambacorta, A., De Rosa, M. & Rossi, M. (1987) Eur. J. Biochem. **167**, 475–479
32. Hanner, M., Redl, B. & Stöffler, G. (1990) Biochim. Biophys. Acta **1033**, 148–153
33. Koch, R., Zablowski, P., Spreinat, A. & Antranikian, G. (1990) FEMS Microbiol. Lett. **71**, 21–26
34. Konishi, J., Wakagi, T., Oshima, T. & Yoshida, M. (1987) J. Biochem. **102**, 1379–1387
35. Lübben, M., Lündsorf, H. & Schäfer, G. (1988) Biol. Chem. Hoppe-Seyler **369**, 1259–1266
36. Dahl, C., Koch, H.-G., Keuken, O. & Truper, H. G. (1990) FEMS Microbiol. Lett. **67**, 27–32
37. Arnone, M. I., Cubellis, M. V., Nitti, G., Sannia, G. & Marino, G. (1988) Ital. J. Biochem. **263**, 12305
38. Löhlein-Werhahn, G., Goepfert, P. & Eggerer, H. (1988) Biol. Chem. Hoppe-Seyler **369**, 109–113
39. Grossebüter, W. & Görisch, H. (1985) Syst. Appl. Microbiol. **6**, 119–124
40. Danson, M. J., Black, S. C., Woodland, D. L. & Wood, P. A. (1985) FEBS Lett. **179**, 120–124
41. Sobek, H. & Gorisch, H. (1988) Biochem. J. **250**, 453–458
42. Smith, L. D., Bungard, S., Danson, M. J. & Hough, D. W. (1987) Biochem. Soc. Trans. **15**, 1097
43. Puchegger, S., Redl, B. & Stöffler, G. (1990) J. Gen. Microbiol. **136**, 1537–1541
44. Costantino, H. R., Brown, S. H. & Kelly, R. M. (1990) J. Bacteriol. **172**, 3654–3660
45. Bragger, J. M., Daniel, R. M., Coolbear, T. & Morgan, H. W. (1989) Appl. Microbiol. Biotech. **31**, 556–561
46. Pisani, F. M., Rella, R., Raia, C. A., Rozzo, C., Nucci, R., Gambacorta, A., De Rosa, M. & Rossi, M. (1990) Eur. J. Biochem. **187**, 321–328
47. Smith, L. D., Budgen, N., Bungard, S. J., Danson, M. J. & Hough, D. W. (1989) Biochem. J. **261**, 973–977
48. Giardina, P., De Baisi, M.-G., De Rosa, M., Gambacorta, A. & Buonocore, V. (1986) Biochem. J. **239**, 517–522
49. Bryant, F. O. & Adams, M. W. (1989) J. Biol. Chem. **264**, 5070–5079
50. Danson, M. J. & Wood, P. A. (1984) FEBS Lett. **172**, 289–293
51. Görisch, H., Hartl, T., Großebüter, W. & Stezowski, J. J. (1985) Biochem. J. **226**, 885–888
52. Guagliardi, A., Moracci, M., Manco, G., Rossi, M. & Bartolucci, S. (1988) Biochim. Biophys. Acta **957**, 301–311
53. Wakao, H., Wakagi, T. & Oshima, T. (1987) J. Biochem. **102**, 255–262
54. Skórko, R., Osipiuk, J. & Stetter, K. O. (1989) J. Bacteriol. **171**, 5162–5164
55. Brown, S. H., Costantino, H. R. & Kelly, R. M. (1990) Appl. Environ. Microbiol. **56**, 1985–1991
56. Prangishivilli, D., Zillig, W., Gierl, A., Biesert, L. & Holz, I. (1982) Eur. J. Biochem. **122**, 471–477
57. Emmel, T., Sand, W., König, W. A. & Bock, E. (1986) J. Gen. Microbiol. **132**, 3415–3420
58. Kikuchi, A. & Asai, K. (1984) Nature (London) **309**, 677–681
59. Barany, F. (1991) Proc. Natl. Acad. Sci. U.S.A. **88**, 189–193
60. Landegren, U., Kaiser, R., Sanders, J. & Hood, L. (1988) Science **241**, 1077–1080
61. Takahashi, M., Yamaguchi, E. & Uchida, T. (1984) J. Biol. Chem. **259**, 10041–10047
62. Whitesides, G. M. & Wong, C.-H. (1985) Angew. Chem. **24**, 617–718

63. Simon, H., Bader, J., Günther, H., Neumann, S. & Thanos, J. (1985) Angew. Chem. **24**, 539–553
64. Jones, J. B. (1987) Enzyme Eng. **8**, 119–128
65. Davies, H. G., Green, R. H., Kelly, D. R. & Roberts, S. M. (1989) Biotransformations in Preparative Organic Chemistry, Academic Press, London,
66. Kovalenko, G. A. & Sokolovskii, V. D. (1983) Biotechnol. Bioeng. **25**, 3177
67. Brink, L. E. S. & Tramper, J. (1985) Biotechnol. Bioeng. **27**, 1258–1269
68. Lamed, R. J., Keinan, E. & Zeikus, J. G. (1981) Enzyme Microb. Technol. **3**, 144–148
69. Keinan, E., Hafeli, E. V., Seth, K. K. & Lamed, R. (1986) J. Am. Chem. Soc. **108**, 162–169
70. Drueckhammer, D. G., Sadozai, S. K. & Wong, C.-H. (1987) Enzyme Microb. Technol. **9**, 564–570
71. Sodano, G., Trabucco, A., De Rosa, M. & Gambacorta, A. (1982) Experientia **38**, 1311–1312
72. Pham, V. T., Phillips, R. S. & Ljungdahl, L. G. (1989) J. Am. Chem. Soc. **111**, 1935–1936
73. Keller, R., Holla, W. & Fülling, G. (1990) U.S. Pat. Appl. 4,963,492
74. Cambou, B. & Klibanov, A. M. (1984) J. Am. Chem. Soc. **106**, 2687–2692
75. Djadchenko, M. A., Pivnitsky, K. K., Thiel, F. & Schick, H. (1989) J. Chem. Soc. Perkin Trans. **1**, 2001–2002
76. Lokotsch, W., Fritsche, K. & Syldatk, C. (1989) Appl. Microbiol. Biotechnol. **31**, 467–472
77. Chien, A., Edgar, D. B. & Trela, J. M. (1976) J. Bacteriol. **127**, 1550–1557

Biochem. Soc. Symp. **58**, 171–180
Printed in Great Britain

Thermoacidophilic archaebacteria: potential applications

P. R. Norris

Department of Biological Sciences, University of Warwick, Coventry CV4 7AL, U.K.

Introduction

The acidophilic archaebacteria that have received most attention during the last twenty years have been assigned to two genera, *Thermoplasma* and *Sulfolobus*, for most of this period. However, it appears that there are many distinct bacteria which superficially resemble *Sulfolobus*, only some of which have so far been distinguished taxonomically. It is the purpose of this article to consider the activity of the *Sulfolobus*-like acidophiles with reference to their diversity and to the interest in their potential industrial utility. The discussion has been limited to areas where acidophily is a key feature of the activity. The potential applications which arise solely through thermophily are not considered beyond noting that several thermostable enzymes from *Sulfolobus* species have attracted attention (for example, the DNA polymerase [1] and β-galactosidase [2]). An amylolytic activity, the capacity for degradation of organosulphur compounds which has been described in the context of bacterial coal desulphurization, and mineral sulphide oxidation in relation to coal desulphurization and the extraction of metals from ores and mineral concentrates are discussed.

The bacteria

Early studies of isolates of *Sulfolobus* revealed strains with different growth temperature optima between 65 and 80 °C [3]. The existence of distinct types among isolates was confirmed by the specificity of immunofluorescence and immunodiffusion reactions [4]. Only one species, *Sulfolobus acidocaldarius*, was designated [5] until other early isolates were distinguished from the type strain and named [6] *Sulfolobus solfataricus* and *Sulfolobus brierleyi* (later to become *Acidianus brierleyi* [7]). Some type-strain cultures have not always conformed to the original descriptions of the bacteria, which has complicated comparisons with new isolates; *S. acidocaldarius* 98-3 (DSM 639), for example, has been found to be unable to oxidize sulphur [8–10]. This

Table 1. Characterized acidophilic, sulphur-oxidizing archaebacteria. In some cases, the indicated mol% G+C values are approximate means of results from several of the given references.

Organism	mol% G+C	Reference
Sulfolobus acidocaldarius	37	[5,6]
Sulfolobus solfataricus	35	[6]
Sulfolobus shibatae	35	[10]
Acidianus brierleyi	31	[7]
Acidianus infernus	31	[7]
Desulfurolobus ambivalens	33	[14]
Metallosphaera sedula	45	[9]

capacity could have been lost by cultures which were originally facultatively autotrophic (strain 98-3 was originally a poor sulphur oxidizer compared with other isolates [11]). Alternatively, some culture lines could now comprise obligate heterotrophs that have been enriched from sulphur-oxidizing mixed cultures as a result of serial culturing in media containing organic nutrients as the carbon and energy sources. Some examples of mixed *Sulfolobus* type-cultures have been noted in a comparison of strains [12]. The use of a relatively straightforward plating procedure with gellan-based solid media allowing growth on sulphur compounds [13] or tryptone and suitable carbohydrates [12] is an improvement on the end-point dilution procedure employed in early isolation work and should facilitate strain purification.

Notwithstanding some anomalous, high values for mol% G+C contents in some of the early descriptions of *Sulfolobus* species, analyses of DNA base contents and DNA:DNA sequence differences of *Sulfolobus*-like bacteria have resulted in the recognition of additional genera (Table 1).

Some isolates have yet to be characterized as fully as those listed (Table 1) but they have nevertheless been examined with regard to their autotrophy and capacity for mineral sulphide oxidation. Strain LM [15–17] is prominent among these and, with the exception of the apparently identical strain BC [18], is readily distinguished from other isolates by comparison of electrophoretic profiles of whole-cell proteins (Fig. 1). However, strain LM/BC remains to be compared with a sulphur- and iron-oxidizing culture line of the *S. acidocaldarius* type-strain. The strains noted in Fig. 1 oxidize sulphur and ferrous iron; of these only *A. brierleyi* (like *Desulfurolobus ambivalens* [14]) has been shown to grow anaerobically with sulphur reduction [7]. Some of these thermoacidophilic archaebacteria can readily switch between autotrophic, mixotrophic and heterotrophic growth [15]. There is also variety in the range of organic substrates which can support wholly heterotrophic growth of different strains of *S. acidocaldarius* and *S. solfataricus* [12]. It would seem worthwhile to identify with certainty the chemolithotrophic and heterotrophic potential of particular strains of the *Sulfolobus*-like bacteria because application could depend on either mode of metabolism; and because a current reference to *S. acidocaldarius*, for example, can lead to expectations of behaviour, based on the original description of the species, that at least certain strains bearing this binomial cannot fulfil.

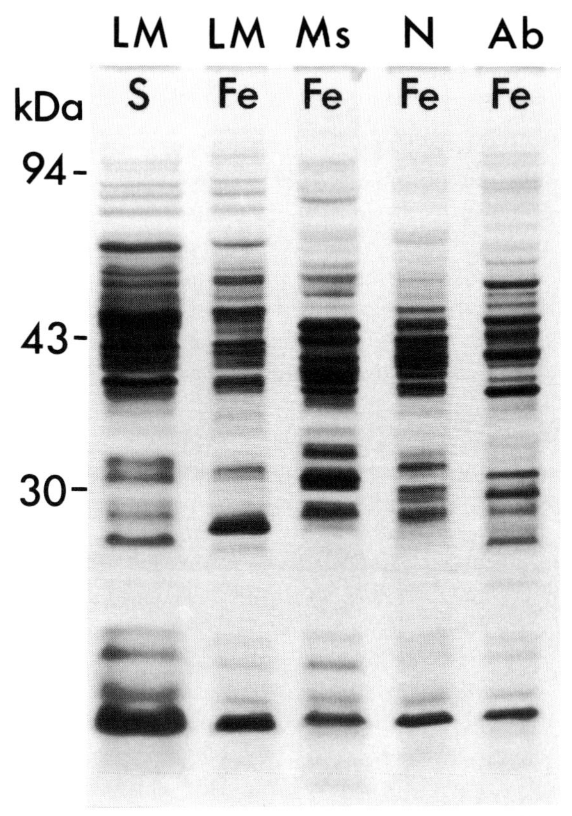

Fig. 1. SDS/PAGE (silver staining) of whole cell proteins of acidophilic archaebacteria. *Sulfolobus* strain LM, *Metallosphaera sedula* (Ms), an uncharacterized isolate from Yellowstone National Park, U.S.A. (N) and *Acidianus brieleyi* (Ab) were grown autotrophically on tetrathionate (S) or ferrous iron (Fe).

Heterotrophic metabolism in application?

There has been little study of the enzymology of the thermoacidophiles in the context of their acid environment, i.e. of the activity of cell surface or extracellularly active enzymes. Isolates of *S. acidocaldarius* and *S. solfataricus* can utilize starch [12]. An amylolytic activity of immobilized *S. solfataricus* has been noted, which has glucose and trehalose as the main products of the starch degradation [19]. In contrast, whole cells of an apparently obligately heterotrophic *Sulfolobus*-like bacterium released maltose, maltotriose and higher glucose polymers before a significant quantity of glucose was observed (P. R. Norris, unpublished work); this strain was received as *S. solfataricus* DSM 1616 but, like *S. acidocaldarius* DSM 639 as discussed earlier, could not oxidize sulphur [8]. The amylolytic activity was inducible and cell associated, with an optimum temperature of about 95 °C for washed cell suspensions, and was greatest at about pH 3. Abiotic starch hydrolysis became significant at pH 2 and superseded that by the bacteria at higher acidity (P. Norris, unpublished work). The enzymic activity was more acid-tolerant but less thermostable than the apparently similar activity of *Pyrococcus woesei* which was optimal at pH 5.5 and observed up to 130 °C [20,21].

S. acidocaldarius is one of many bacteria investigated as being potentially useful for coal desulphurization (see [22] for review) but is so far unique in that the possible application of both its heterotrophic and chemolithotrophic metabolic activities has been raised. The removal of the inorganic sulphur fraction of coal through the bacterial oxidation of pyrite (FeS_2) has been demonstrated numerous times in the laboratory but the problems of developing an economic process on the massive scale required for widespread application remain formidable. The great majority of the laboratory work has used *Thiobacillus ferrooxidans* or mixed cultures containing this mesophilic, acidophilic eubacterium. The attraction of using the thermoacidophilic archaebacteria lies primarily in their more rapid oxidation and removal of the pyrite [23]. The interest in the activity of *S. acidocaldarius* for removing organic sulphur from coal stems from reports [24,25] describing its capacity to degrade some of the thiophene compounds which may be found in the coal matrix and which represent a more difficult target for removal than the discrete iron sulphide particles. The examination of organosulphur compound degradation in this context has generally involved neutrophilic, heterotrophic mesophiles. This would imply a two-stage process for removal of both inorganic sulphur (by mineral sulphide-oxidizing acidophiles) and organic sulphur (by heterotrophic neutrophiles).

Thiophenes have been provided as sole carbon and energy sources for *S. acidocaldarius* which was previously grown on glucose; the gradual accumulation of sulphate in the medium was described but only a slight increase in cell number was noted [24,25]. In other work, dibenzothiophene has appeared unstable at pH 3 and 70 °C [26]; the rate of its disappearance was not enhanced by the presence of the obligately heterotrophic thermoacidophile which was received as *S. solfataricus* and exhibited the amylolytic activity described earlier. However, thiophene-2-carboxylate was apparently metabolized by this thermoacidophile; the oxygen consumption of washed cell suspensions being stimulated by about half as much as by succinate. The thiophene did not serve as a sole carbon and energy source for growth but the rate of its oxidation was doubled by prior exposure to low concentrations of thiophene during growth on yeast extract [26]. Thiophene-2-carboxylate did not stimulate oxygen uptake by the sulphur-oxidizing strains *A. brierleyi* and *Sulfolobus* B6-2, even when they were grown heterotrophically in its presence.

Biohydrometallurgy

The lithoautotrophic metabolism of certain thermoacidophiles is central to their potential application in bacterial leaching and mineral processing, as well as in the removal of pyrite from coal. The interest in the activity of mineral sulphide-oxidizing thermophiles was prompted, at least in part, by the observations of high temperatures in some low-grade ore leach dumps where the bulk of the material prevents the dissipation of heat generated by exothermic oxidation of mineral sulphides. Brierley & Brierley [27] have reviewed the potential application of thermophiles in ore leaching and noted one experiment in which a mass of chalcopyrite ore (5.8 tonnes)

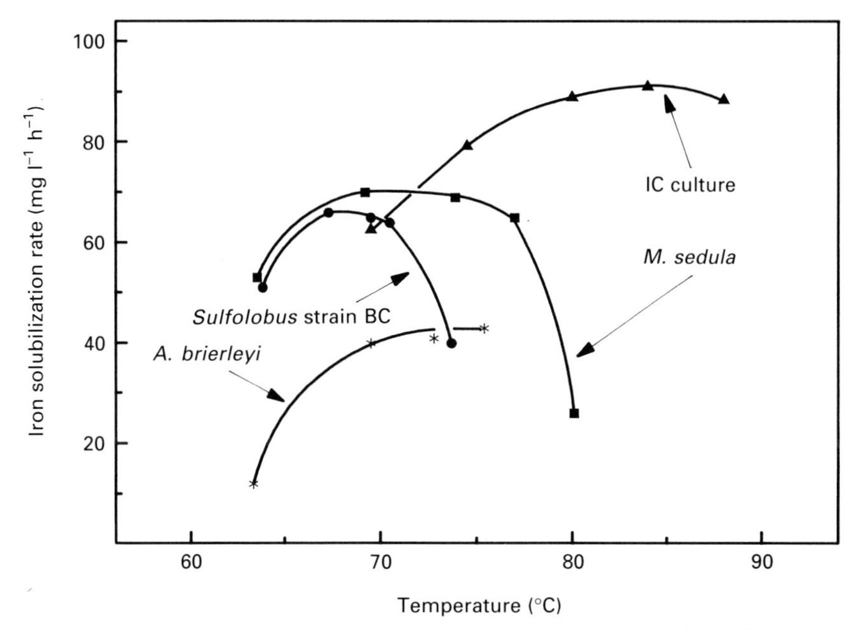

Fig. 2. The effect of temperature on pyrite dissolution during growth of acidophilic archaebacteria in air-lift reactors. The IC culture is an uncharacterized *Sulfolobus*-like bacterium. A simple mineral salts medium containing pyrite (40% Fe; 1% (w/v) < 75 μm diameter particles) was gassed with 1% (v/v) CO_2 in air.

was inoculated with *Sulfolobus*: the rate of copper solubilization was over six-fold greater than that previously obtained at 25 °C, but was not maintained beyond 120 days. It is easier, with regard to more reproducible experimentation and to the diversity of the *Sulfolobus*-like bacteria, to consider the potential application of the extreme thermophiles for mineral sulphide processing in reactors.

Currently, the industrial use of bioreactors is restricted to the release of gold from mineral sulphides [28] by mesophilic eubacteria (principally *Thiobacillus ferrooxidans*). However, higher temperatures (about 45 °C) have been used in a successful pilot plant demonstration utilizing moderately thermophilic eubacteria [29]. The advantage of using thermophilic archaebacteria in the treatment of refractory gold ores and flotation concentrates at even higher temperatures (60 °C) has been demonstrated in the laboratory [30]. The potential benefits of using the thermophiles include the alleviation of reactor cooling costs incurred by operating at lower temperatures and the more rapid mineral oxidation that can be obtained. In addition, the extent of degradation of some mineral sulphides is increased at elevated temperatures resulting in higher yields of the target metal in solution. The incomplete oxidation of chalcopyrite ($CuFeS_2$) by bacteria at low temperatures arises through the deposition of secondary minerals and precipitates at the mineral surface; the increased copper yield obtainable with *Sulfolobus*-like bacteria (strain LM/BC) at 70 °C has been well documented [8,31,32]. Alternative but potentially expensive strategies for achieving efficient copper extraction with mesophiles have included

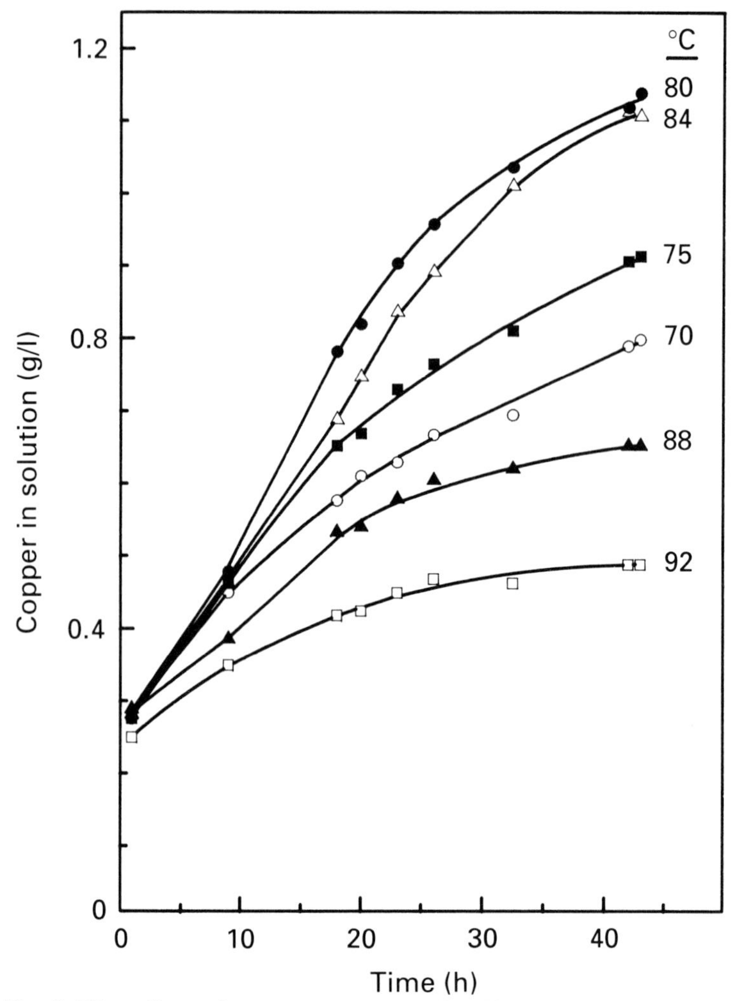

Fig. 3. The effect of temperature on the dissolution of a chalcopyrite concentrate [20% Cu; 1% (w/v)] by *Sulfolobus*-like bacteria. The IC culture and conditions are described in Fig. 2.

periodic regrinding of the mineral particles to expose fresh surfaces [33] and the use of silver as a catalyst to disrupt the passivation layer [34]. *Sulfolobus*-like bacteria which are active at higher temperatures and over a broader temperature range than the well-studied strain LM/BC promote more rapid and efficient mineral sulphide oxidation (Figs. 2 and 3) and their use would require less exacting process control with regard to the temperature.

The potential toxicity of metal ions at the high concentrations that could be reached in bioreactors is one of several factors that would have to be considered in the selection of an appropriate acidophilic archaebacterium for mineral processing. *Sulfolobus* strain BC has been shown to tolerate between 25 and 30 g of copper/l during chalcopyrite oxidation [16,32] whereas *Metallosphaera sedula* [9] is inhibited by

Table 2. Characteristics of iron oxidation by acidophilic bacteria ([35]; P. R. Norris, unpublished work). Doubling times given are for iron oxidation during exponential, autotrophic growth. The initial ferrous sulphate concentration was 50 mM at pH 1.6, 68 °C for the archaebacteria [cultures gassed with 1% (v/v) CO_2 in air] and 30 °C for the mesophiles. Oxidation of 0.15 mM-ferrous sulphate/h that occurred in sterile controls at 68 °C was subtracted from the oxidation rate in the thermophile cultures to give the doubling times. Abbreviations used: n.d., not determined; un-c, uncompetitive inhibition.

Organism	Doubling time (h)	V_{max}[a]	K_m (mM)[a]	K_i, Fe^{3+} (mM)[a]	K_i, Cu (mM)
Sulfolobus strain BC	8.5	2.7	0.4	1.7	58
Metallosphaerea sedula	5.0	3.2	1.0	1.0	n.d.
Acidianus brierleyi	9.5	1.1	0.4	1.9	n.d.
Thiobacillus ferrooxidans	6.5	1.6–3.0	1.3	3.1	230
Leptospirillum ferrooxidans	11.0	0.5–1.4	0.25	42.8	un-c[b]

[a] Kinetic data for ferrous iron oxidation were obtained with washed cell suspensions and oxygen electrodes. $V_{max.}$ units are μmol O_2 uptake min^{-1} mg^{-1} of protein.
[b] see [35].

about 5 g/l and the IC culture that grows well at 85 °C (see Fig. 2) is inhibited by 10 g/l (J. Owen & P. Norris, unpublished work).

The oxidation of ferrous iron is probably the key reaction in dissolution of the industrially important mineral sulphides by acidophiles. Different characteristics of iron oxidation among different bacteria, reflecting different oxidation systems, can provide a basis for strain selection with regard to mineral oxidation capacities. This has been discussed for mesophilic and moderately thermophilic iron-oxidizing eubacteria [31,35], but how and which factors influence iron oxidation and the process itself in the archaebacteria has received little study. This is partly because of the problems of significant spontaneous iron oxidation at high temperature, except towards the lower end of the pH range tolerated by the bacteria, and because of the problems of accurately measuring the biomass under conditions where the product, ferric iron, tends to precipitate. A summary of preliminary results for iron oxidation by archaebacteria is presented in comparison with data for the relatively well-studied mesophilic, acidophilic eubacteria, *Thiobacillus ferro-oxidans* and *Leptospirillum ferro-oxidans*; the iron being supplied as ferrous sulphate in solution rather than as a mineral sulphide (Table 2).

There are differences in growth rates and iron oxidation kinetics among the thermophiles but the general levels of activity are similar to those of eubacteria. In the absence of inherently faster growth of the thermophiles, their higher rate of mineral sulphide oxidation, in comparison with that of mesophiles, presumably results from the influence of the temperature on overall rate-limiting reactions (such as ion diffusion) at the solid substrate surface or through superficial reaction layers.

The respiratory chains involved in the transfer of electrons from ferrous iron appear to differ in phylogenetically distinct iron-oxidizing acidophiles [36]. Within the archaebacteria, however, an identical novel cytochrome is induced in *Sulfolobus* strain LM/BC, *M. sedula*, *A. brierleyi* and other less-well-described iron-oxidizing isolates during growth on ferrous iron ([36]; P. R. Norris, unpublished work). Initial experiments have indicated that this cytochrome, with a characteristic α-absorption peak at 572 nm in reduced −oxidized difference spectra, is membrane-bound and has an appropriate midpoint potential at pH 3.5 for a role in electron transfer from ferrous iron (A. Hart & P. R. Norris, unpublished work). A cytochrome with an absorption spectrum characteristic of the *S. acidocaldarius* cytochrome aa_3 [37] is present simultaneously with the novel cytochrome in the iron-oxidizing isolates. Even if the respiratory chains of the various iron-oxidizing archaebacteria prove to be similar, some strain-specific features of the interaction with the substrate are indicated by the differences in the iron oxidation kinetics (Table 2). SDS/PAGE of whole cell proteins of *Sulfolobus* strain LM grown on different substrates appears to show increased production of a protein, in addition to the novel cytochrome, during growth on iron (see Fig. 1). The polypeptide with an apparent molecular mass of about 27 kDa in the denaturing gel was mostly solubilized on cell breakage in contrast with the novel cytochrome, although some remained with the membrane fraction (A. Hart & P. Norris, unpublished work). The nature of the protein, any direct role for it in iron oxidation, and whether it is produced by the other isolates as well as strain LM, all remain to be elucidated.

In addition to an adequate capacity for iron and sulphur oxidation and to a sufficient tolerance of potentially toxic metal ions, the bacteria would also have to withstand agitation in the presence of high concentrations of mineral particles to be worthy of consideration for application in bioreactors. Considerable diversity in 'robustness' has been shown by different isolates (P. Norris, unpublished work) and future work in this context could investigate whether the range of iron-oxidizing isolates show contrasting stabilities of S-layers as is the case for *S. acidocaldarius* and *A. brierleyi* [38].

Conclusions

The biochemical reactions on which potential applications of the acidophilic archaebacteria would depend have received little attention. Thiophene degradation and iron oxidation by *Sulfolobus*-like bacteria are poorly characterized, and it remains to be established unequivocally whether the capacities for degrading organosulphur compounds and pyrite are possessed by a single species. Desirable characteristics of thermoacidophiles which could be considered for application in the extraction of metals from mineral sulphides, such as a broad temperature range for activity and a high tolerance of potentially toxic metal ions, have been found in different strains. The taxonomic characterization of several so far un-named isolates and the collection of further novel strains should allow the identification of those with appropriate combinations of the desirable features and will continue to widen the perceived diversity of the acidophilic, *Sulfolobus*-like bacteria.

Original work described in this article has been supported by the S.E.R.C. Biotechnology Directorate and by CRA Ltd. Advanced Technical Development, Australia.

References

1. Elie, C., De Recondo, A. M. & Forterre, P. (1989) Eur. J. Biochem. **178**, 619–626
2. Pisani, M., Rella, R., Raia, C. A., Rozzo, C., Nucci, R., Gambacorta, A., De Rosa, M. & Rossi, M. (1990) Eur. J. Biochem. **187**, 321–328
3. Mosser, J. L., Mosser, A. G. & Brock, T. D. (1974) Arch. Microbiol. **97**, 169–179
4. Bohlool, B. B. & Brock, T. D. (1974) Arch. Microbiol. **97**, 181–194
5. Brock, T. D., Brock, K. M., Belly, R. T. & Weiss, R. L. (1972) Arch. Mikrobiol. **84**, 54–68
6. Zillig, W., Stetter, K. O., Wunderl, S., Schulz, W., Priess, H. & Scholz, I. (1980) Arch. Microbiol. **125**, 259–269
7. Segerer, A., Neuner, A., Kristjansson, J. K. & Stetter, K. O. (1986) Int. J. Syst. Bacteriol. **36**, 559–564
8. Marsh, R. M., Norris, P. R. & Le Roux, N. W. (1983) in Recent Progress in Biohydrometallurgy (Rossi, G. & Torma, A. E., eds.), pp. 71–81, Associazione Mineraria Sarda, Iglesias
9. Huber, G., Spinnler, C., Gambacorta, A. & Stetter, K. O. (1989) Syst. Appl. Microbiol. **12**, 38–47
10. Grogan, D., Palm, P. & Zillig, W. (1990) Arch. Microbiol. **154**, 594–599
11. Shivvers, D. W. & Brock, T. D. (1973) J. Bacteriol. **114**, 706–710
12. Grogan, D. (1989) J. Bacteriol. **171**, 6710–6719
13. Lindström, E. B. & Sehlin, H. M. (1989) Appl. Environ. Microbiol. **55**, 3020–3021
14. Zillig, W., Yeats, S., Holz, I., Böck, A., Rettenberger, M., Gropp, F. & Simon, G. (1986) Syst. Appl. Microbiol. **8**, 197–203
15. Wood, A. P., Kelly, D. P. & Norris, P. R. (1987) Arch. Microbiol. **146**, 382–389
16. Norris, P. R. & Parrott, L. (1986) in Fundamental and Applied Biohydrometallurgy (Lawrence, R. W., Branion, R. M. R. & Ebner, H. G., eds.), pp. 355–365, Elsevier, Amsterdam
17. Norris, P., Nixon, A. & Hart, A. (1989) in Microbiology of Extreme Environments and its Potential for Biotechnology (Da Costa, M. S., Duarte, J. C. & Williams, R. A. D., eds.), pp. 24–43, Elsevier, London
18. Norris, P. R., Marsh, R. M. & Lindström, E. B. (1986) Biotechnol. Appl. Biochem. **8**, 318–329
19. Lama, L., Nicolaus, B., Trincone, A., Morzillo, P., De Rosa, M. & Gambacorta, A. (1990) Biotechnol. Lett. **12**, 431–432
20. Brown, S. H., Costantino, H. R. & Kelly, R. M. (1990) Appl. Environ. Microbiol. **56**, 1985–1991
21. Koch, R., Spreinat, A., Lemke, K. & Antranikian, G. (1991) Arch. Microbiol. **155**, 572–578
22. Bos, P. & Kuenen, J. G. (1990) in Microbial Mineral Recovery (Ehrlich, H. L. & Brierley, C. L., eds.), pp. 343–377, McGraw-Hill, New York
23. Detz, C. M. & Barvinchak, G. (1979) Min. Congr. J. **65**, 75–82
24. Kargi, F. & Robinson, J. M. (1984) Biotechnol. Bioeng. **26**, 687–690
25. Kargi, F. (1987) Biotechnol. Lett. **9**, 478–482
26. Constanti, M., Giralt, J., Bordons, A. & Norris, P. R. (1992) Appl. Biochem. Biotechnol. **34/35**, in the press
27. Brierley, J. A. & Brierley, C. L. (1986) in Thermophiles: General, Molecular, and Applied Microbiology (Brock, T. D., ed.), pp. 279–305, Wiley, New York

28. van Aswegen, P. C., Haines, A. K. & Marais, H. J. (1988) in Randol Perth Gold **88** (Conf. Proc.), pp. 144–147, Randol, U.S.A.
29. Spencer, P. A., Budden, J. R. & Sneyd, R. (1989) in Biohydrometallurgy 1989 (Salley, J., McCready, R. G. L. & Wichlacz, P., eds.), pp. 231–242, Canmet, Ontario
30. Hutchins, S. R., Brierley, J. A. & Brierley, C. L. (1987) in Process Mineralogy VII (Vassiliou, A. H., Hausen, D. M. & Carson, D. J. T., eds.), pp. 53–66, The Metallurgical Society, Warrendale, U.S.A.
31. Norris, P. R. (1990) in Microbial Mineral Recovery (Ehrlich, H. L. & Brierley, C. L., eds.), pp. 3–27, McGraw-Hill, New York
32. Le Roux, N. W. & Wakerley, D. S. (1988) in Biohydrometall. Proc. Int. Symp. (Norris, P. R. & Kelly, D. P., eds.), pp. 305–317, Science and Technology Letters, Kew
33. McElroy, R. O. & Bruynesteyn, A. (1978) in Metallurgical Applications of Bacterial Leaching and Related Microbiological Phenomena (Murr, L. E., Torma, A. E. & Brierley, J. A., eds.), pp. 441–462, Academic Press, New York
34. Bruynesteyn, A., Lawrence, R. W., Vizsolyi, A. & Hackl, R. (1983) in Recent Progress in Biohydrometallurgy (Rossi, G. & Torma, A. E., eds.), pp. 151–168, Associazione Mineraria Sarda, Iglesias
35. Norris, P. R., Barr, D. W. & Hinson, D. (1988) in Biohydrometall. Proc. Int. Symp. (Norris, P. R. & Kelly, D. P., eds.), pp. 43–59, Science and Technology Letters, Kew
36. Barr, D. W., Ingledew, W. J. & Norris, P. R. (1990) FEMS Microbiol. Lett. **70**, 85–95
37. Anemüller, S. & Schäfer, G. (1989) Eur. J. Biochem. **191**, 297–305
38. König, H. & Stetter, K. O. (1986) Syst. Appl. Microbiol. **7**, 300–309

Biochem. Soc. Symp. **58**, 181–193
Printed in Great Britain

Biotechnological potential of methanogens

Lacy Daniels

Department of Microbiology, University of Iowa, Iowa City, IA 55242, U.S.A.

Synopsis

Methane produced microbiologically is currently used as an energy source, especially by cities and industries, albeit at a level far below its potential; the incentive is currently to save money on disposal costs for waste problems. Anaerobic digestion can be helpful in degrading several halogenated hydrocarbon wastes, and methanogens are partly responsible. Ethane instead of methane may be a future product of interest. Some pure cultures of methanogens may be suitable for production of B-12, or perhaps the speciality biochemical F_{420}, a 5-deazaflavin of interest to both methanogen and streptomyces researchers.

Methanogens can cause a variety of problems, including biocorrosion, increased atmospheric methane, and ruminant nutrition loss. Studies of tropical wetlands, including rice paddies and swamps, and the study of a variety of ruminants in the tropics are particularly interesting and appropriate at this time, with respect to methane produced in these ecosystems. In some cases, it may be possible to control methane production by the use of inhibitors or ecological control mechanisms.

Introduction

Methanogenic bacteria are a diverse group of archaebacteria that share the ability to make methane as a means of obtaining energy [1–3]. Although they share some properties with other archaebacteria, they also have biochemical and microbiological features unique for their group; for example, several of the coenzymes used in the methanogenic pathway are not found in large quantities or at all in non-methanogens [4]. Other groups of archaebacteria are restricted to growth only at mesophilic (e.g. halobacteria), or at thermophilic (e.g. sulphur-dependent thermophiles) temperatures, but methanogens as a group cover a wide range of possible growth temperatures; some grow well at 20 °C, and others can grow at 100 °C [3,5]; although methane production at low temperatures (5–15 °C) has been observed, no psychrotrophic or psychrophilic strains have been isolated ([3,5–8];

K. Y. Jung & L. Daniels, unpublished work). Since anaerobic conditions are required for growth, they are commonly found in anaerobic ecosystems with suitable substrates, namely hydrogen, carbon dioxide, formate, acetate, methanol, or methylamines; these substrates arise because of the activities of a variety of bacteria that degrade complex materials, such as cellulose, starch, and proteins. Thus, methanogens will be found in most anaerobic muds at lake bottoms, rivers, estuaries, and oceans, over a temperature range of 20–100 °C. They are also found in human and animal guts and rumens, and are an important component of anaerobic waste treatment systems [3,5,9–11].

The potential usefulness of methanogens is controlled by the properties mentioned above, as are any problems they cause. Since they must make methane to live, we can use this product as a source of energy. In some cases, we can make use of their coenzymes, both those that are unusual and any that are present at high levels (e.g. 5-deazaflavins and corrinoids) [4]. It also may be possible to modify methanogens to make alternative products, e.g. ethane instead of methane [12]. In addition to their positive side, in some cases methanogens cause problems. Methanogens participate in anaerobic biocorrosion [13]. Their ability to produce methane is a double-edged sword, since it is a 'greenhouse' gas, and its release into the atmosphere will probably lead to global warming problems; it is increasing in our atmosphere at an alarming rate, due to methanogenic activity more than from geological sources [14]. Also, methane release from ruminants represents energy lost from the animal's diet [15]. Thus, in addition to harnessing methanogen products, we must consider how to control their growth in certain problem environments. To do so, we must draw upon information from several disciplines, including biochemistry, microbiology, and ecology of methanogens. In this chapter, I describe the biotechnological potential for the productive use of methanogens to make methane and ethane, their degradation of some halogenated wastes, and their production of corrinoids and 5-deazaflavins as speciality biochemicals; I also discuss problems caused by methanogens in biocorrosion, global warming, and ruminant nutrition, and mention several ways methanogens may be controlled via inhibitors or ecological control mechanisms.

Methane production for use as an alternative source of energy

Methane production for use as an energy source is a biotechnological potential that, on a small scale, is already a reality. Some cities and industries currently use anaerobic digestion to deal with their waste problems, and its use has gained favour during the past decade as an alternative to aerobic treatment processes. A variety of papers and reviews have detailed the promise and advantages of this form of treatment, including its use to provide energy [10,11,16–22], and several countries (notably India and China) have undertaken massive small-scale methane production projects to use agricultural and rural domestic wastes [16,17]. Despite the apparent advantages of producing energy from wastes, the economics of such processes are seldom favourable; in general, small facilities have not worked well, and it is currently thought that rural production of methane on small farms will not be either cost effective or routinely suitable even as a process for waste treatment, because of the time and expertise required to run an effective system. Thus, the massive projects in India and China have not been very successful, unless they were part of a community system.

In the U.S.A. and Europe, the most successful examples are found either in industries (especially food processing industries) or in municipal wastewater treatment. One key factor in their success has been the introduction of new process configurations, all of which retain biomass to some extent. The major driving force has been that the energy obtained can be used to off-set the costs of waste treatment that must be done, and that the process is effective and inexpensive, compared with aerobic treatment; the trend toward anaerobic processes has seldom been due to the altruistic desire to save energy or to make money selling methane as a fuel. Fig. 1 describes several types of reactors that help retain bacteria that are produced in the wastes, so as a result the catalyst is not being lost as the waste stream moves along [10,11,16–22]. The first improvement to the traditional stirred reactor was an added settling system that recycled the biomass. Both the expanded-bed and stationary-support designs create a 'fixed film' of bacteria that remain behind, and are also very resistant to toxic materials, and accidental or intentional shutdowns. The UASB (upflow anaerobic sludge blanket) retains cells as a sludge of biomass granules. Combinations and variations of these designs, and their use in series are seen in some cases. A very different sort of 'reactor' is found in the traditional landfill, with alternate layers of soil and municipal wastes, where a collection system can be built to harvest the methane. Examples of the waste types being treated are given in Table 1; some of these food industry reactors are at a very large scale, e.g. 1–10 million litres capacity, with municipal wastewater reactors considerably larger. However, in many cases the methane produced is not productively used, or in some cases a portion is used to blend with commercially purchased natural gas, in an effort to avoid treatment of the gas to remove sulphide and carbon dioxide; in many cases, it is cheaper to buy natural gas than to make the capital investment for a gas purification system.

As the prices of petroleum-based fuels increase in the coming decades, more attention will be paid to biologically produced methane. The anaerobic process has several advantages besides the generation of methane—there are lower operational energy costs (partly due to the absence of air pumping costs that accompany aerobic systems), and also less waste sludge is produced (since growth yields during anaerobic growth are less than with aerobic growth). As this economic transition occurs, it will be worthwhile examining inexpensive and abundant organic feedstocks (e.g. starch or cellulose) as large-scale agricultural products to be used specifically for methane production. The use of agricultural wastes, e.g. crop residue, may be attractive at first glance, but unattractive in reality, because of the energy and expense needed to gather and transport it to a suitable large-scale facility. Another possible feedstock for methane production is syngas, a mixture of hydrogen, carbon monoxide, and carbon dioxide that is created by controlled combustion of coal [10,23].

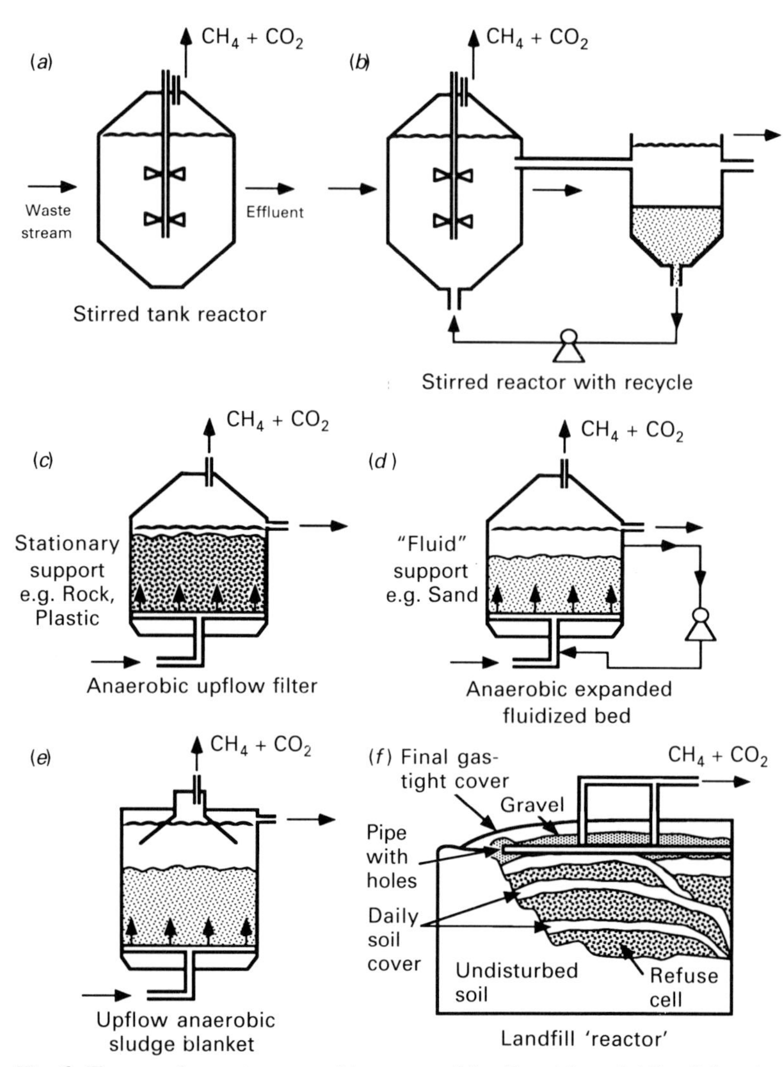

Fig. 1. Types of reactors used in anaerobic digestion. *(a)* Traditional stirred-tank reactor, now seldom used alone in process design. *(b)* Traditional stirred-tank reactor with addition of settling tank for cell recycle. (*c*) Stationary support with cells growing on fixed support material, e.g. plastic rings or stones. (*d*) Expanded-bed design, with cells growing on particulate media (e.g. sand) which is kept suspended by recirculating liquid. (e), Upflow anaerobic sludge blanket (UASB), with cells settling as granules. (*f*) Landfill, with alternating soil and waste layers, and a capped top with a methane recovery system. See also [10,11,16–22].

Table 1. Examples of anaerobic waste treatment. See Henze & Harremoes [22] and Daniels [10].

Type of waste	kg COD/m^3	Process type
Rum distillery	55	Fixed-bed
Bottling	7	Fluidized-bed
Potato chip	85	Recycled flocs
Sugar beet	3–17	Sludge blanket
Domestic	–	Landfill

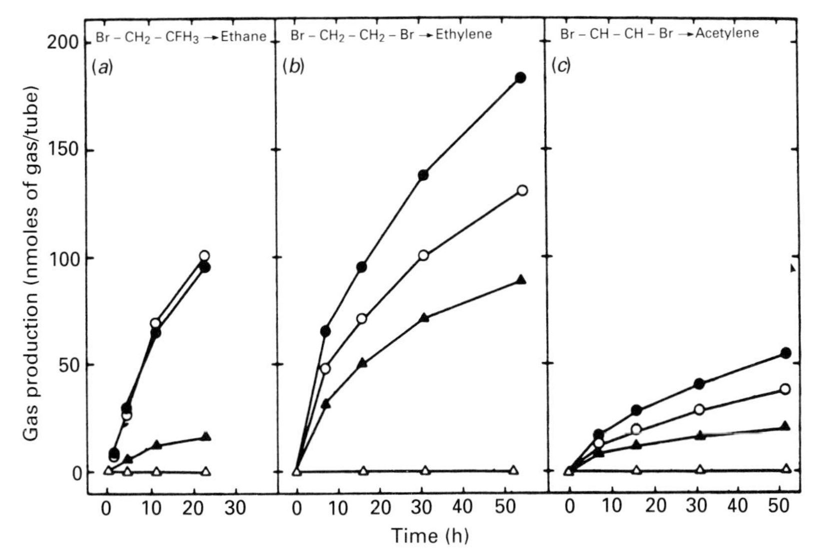

Fig. 2. Effects of halogenated hydrocarbon concentrations on the production of 2-carbon gases by mid-logarithmic-phase cultures of methanogens. (*a*) Ethane production by *Methanococcus thermolithotrophicus* in the presence of various concentrations of bromoethane: ○, 119 μM; ●, 60 μM; ▲, 24 μM; △, control. (*b*) Ethylene production by the same organism in the presence of dibromoethane: ●, 213 μM; ○, 108 μM; ▲, 23 μM; △, control. (*c*) Acetylene production by *Methanobacterium thermoautotrophicum* in the presence of dibromoethylene: ●, 1000 μM; ○, 509 μM; ▲, 210 μM; △, control. Reprinted with permission from Belay & Daniels [27].

Anaerobic treatment of halogenated hydrocarbon wastes

Several laboratories have reported that some halogenated alkane hydrocarbons can be degraded by microbes found in anaerobic digestors [24–29]. In some cases pure cultures of methanogens can convert simple haloalkanes into harmless gases, ethane, ethylene, or acetylene [27]; Fig. 2 provides an example of this, using *Methanococcus* and *Methanobacterium* species. The level of methane production is

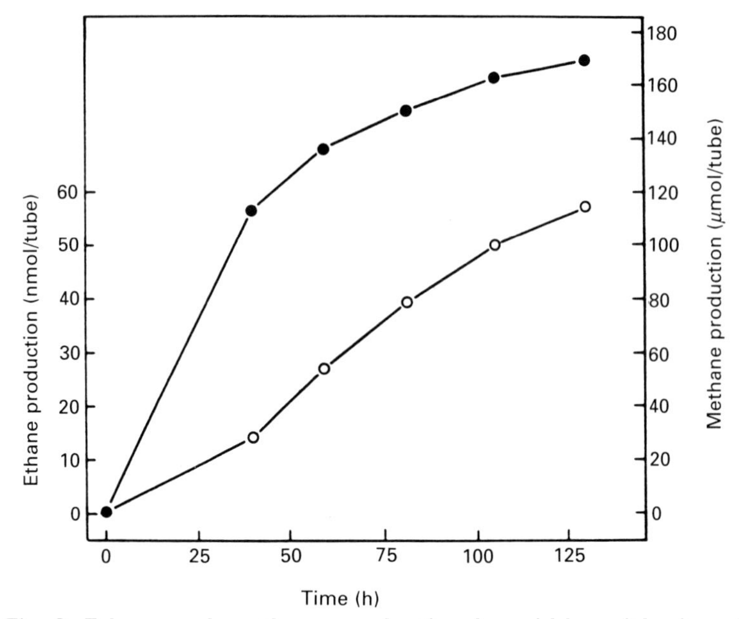

Fig. 3. Ethane and methane production by mid-logarithmic cultures of *Methanosarcina barkeri* strain 227 supplemented with 5.3% (v/v) ethanol in hydrogen plus carbon dioxide medium. Key to symbols: ○, ethane; ●, methane. Medium and all conditions were as described previously by Belay & Daniels, [12]. The values are for tubes of 25.7 ml capacity with 5.6 ml of culture.

typically about 1000-fold higher than the two-carbon gases, unless enough halogenated material is present to inhibit methanogenesis. It is thought that corrinoids are involved in many of these reactions, and a variety of non-methanogenic anaerobes may be responsible for similar reactions. Many recalcitrant wastes may be best treated by a sequence of anaerobic and aerobic digestion steps.

Ethane production from ethanol

Some strains of *Methanosarcina barkeri* produce ethane when grown in the presence of ethanol [12]. Fig. 3 describes ethane production during growth of a *Methanosarcina* strain on H_2–CO_2. Note that ethane production is much lower than methane production; typically ethane levels do not exceed 1 % in the gas phase. Ethane production is dependent on ethanol level; generally the optimum is at 1–5 % ethanol, depending on the strain.

Ethane production is of interest in several ways. Ethane is a higher-value gas than methane; because of its higher volumetric energy content and compressibility it is preferred for some applications; it is also a potential feedstock for ethylene production, and thus polymer production. Although it is not known what pathway

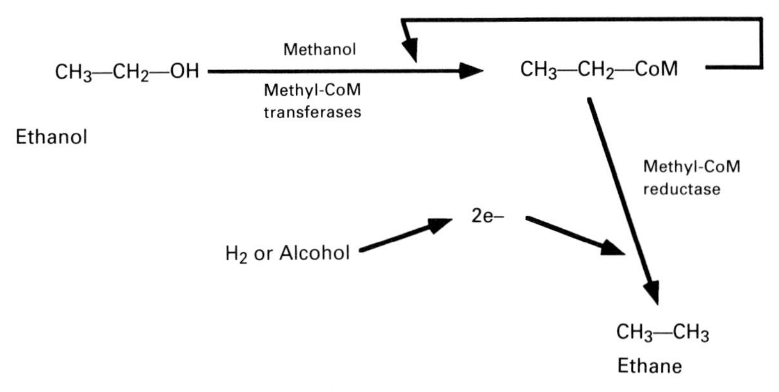

Fig. 4. Overview of putative steps in the conversion of ethanol into ethane by *Methanosarcina* species. This path is in analogy to the pathway of methanol conversion into methane [1,2].

is used to convert the ethanol into ethane, it may use the same pathway as that used for methane production from methanol plus hydrogen, as shown in reactions **1** and **2**, and in Fig. 4:

$$CH_3OH + H_2 \rightarrow CH_4 + H_2O \quad (1)$$

$$CH_3{-}CH_2{-}OH + H_2 \rightarrow CH_3{-}CH_3 + H_2O \quad (2)$$

This implies the methanogenic enzymes involved in this path can use two-carbon compounds as alternative substrates. Fig. 4 shows an overview of possible steps in the production of ethane from ethanol, using either hydrogen or methanol as electron donors. Two methyltransferases (MTs) are probably involved, MT-1 and MT-2, which were originally described by the Vogels group in Nijmegen [30–33]; however, their use in transferring ethyl groups has not yet been demonstrated. MT-1 is an oxygen-sensitive corrinoid-containing protein, and MT-2 is an oxygen-stable enzyme which transfers the methyl (and possibly the ethyl) group onto coenzyme M (CoM). The terminal reaction is catalysed by methyl-CoM reductase [2,34–36], a complex enzyme system using two electrons to reduce the alkyl group to produce methane; this enzyme (from *Methanobacterium thermoautotrophicum*) has been shown in two studies to use ethyl-CoM as an alternate substrate [36,37].

Methanogen-produced coenzymes as potential products

Two methanogen coenzymes deserve mention as potential products: corrinoids, and the 5-deazaflavin electron-transfer coenzyme known as F_{420}. Structures of both are shown in Fig. 5.

Work on the potential of methanogens to produce industrial quantities of corrinoids has been conducted mainly in Japan, in the laboratories of Nagai, Mazumder, Nishio, and others [38,39]. Fig. 5 describes the structural distinctions between the ‘normal’ cyanocobalamin (B-12; the 5,6-dimethylbenzimidazole derivative), and two structural variations found in methanogens, B-12-HBI (the

Table 2. Levels of corrinoids produced by bacteria.

Organism	Corrinoid concn. (mg/l)	Reference
Propionibacterium (industry)	23	[44]
Pseudomonas (industry)	60	[44]
Butyribacterium	93	[43]
Clostridium	6	[40]
Methanosarcina	8	[38,39]

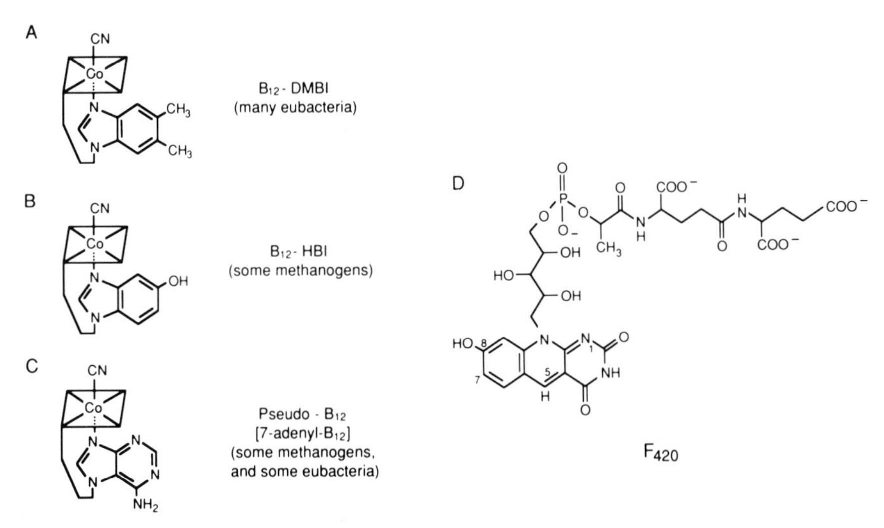

Fig. 5. Structures of corrinoids and F_{420} (5-deazaflavin) found in methanogens. A, B_{12}-DMBI (dimethylbenzimidazole); B, B_{12}-HBI (hydroxybenzimidazole); C, pseudo-B_{12} (7-adenyl); D, F_{420}.

hydroxybenzimidazole derivative) and pseudo-B-12 (the 7-adenyl derivative) [2,4,40–42]; work on corrinoids in methanogens has concentrated on *Methosarcina*, which produces relatively high levels of B-12-HBI, compared with other methanogens [38–40]. Table 2 describes the levels of corrinoids produced by several bacteria; although the levels of B-12 produced by methanogens are low compared with the industrially used strains, production has not yet been optimized by mutation or genetic engineering [38–44]. However, given the relative difficulty and cost of growing methanogens compared with the industrially used strains, it is likely that non-methanogens will retain their domination of the market. Alternatively, methanogen biomass as a by-product of another process may provide a value-added product, e.g. for animal feed.

The 5-deazaflavin F_{420} is a major coenzyme in methanogens [1,2,4], and acts as electron acceptor during the oxidation of hydrogen and formate, and as an electron donor in several steps of the methanogenic pathway [see Fig. 5]. It is present in methanogens at about 0.1–3 mg/g of dry cell weight. F_{420} is also present in all the *Streptomyces* strains thus far examined [45,46], but at a level about 10–100-fold lower than methanogens. Its role in most streptomyces is not known, but in at least three instances, F_{420} is an electron donor in the synthesis of an antibiotic; the biosynthesis of 7-chlorotetracycline, oxytetracycline, and lincomycin are dependent on this coenzyme [47–49]. It also serves as an electron transfer cofactor in the NADP-F_{420} oxidoreductase in several streptomyces (L. Daniels, D. Lambert & K. Harmon, unpublished work). Currently, F_{420} is not commercially available; an abbreviated structure, F_0, can be synthesized chemically [50], but this form exhibits some kinetic properties different from the native coenzyme [4,51]. Most laboratories studying methanogens prepare F_{420} by growing methanogens, extracting the coenzyme, and purifying it by column chromatography. We have recently completed a study of how to most effectively grow methanogens to a high yield, extract, and purify the coenzyme (E. Purwantini, B. Mukhopadhyay & L. Daniels, unpublished work); we conclude our method will give improved and consistent yields, and will avoid the formation of degradation products probably found in some reported preparations. Given the potential value of F_{420} in antibiotic research and even production, and its value in methanogen biochemical research, it may represent a significant, although small-scale, speciality product available from methanogens.

Methanogens as problems

Biocorrosion

We have proven methanogens can make use of electrons derived from elemental iron to produce methane from carbon dioxide [13]. They do so by rapidly consuming the very low levels of free hydrogen produced chemically, and thus make the overall reaction **3** energetically favourable ($\Delta G0' = -136$ kJ per reaction):

$$8H^+ + 4Fe^\circ + CO_2 \rightarrow CH_4 + 4Fe^{2+} + 2H_2O \qquad (\mathbf{3})$$

This worked proved that the theory of cathodic depolarization as a mechanism of anaerobic biocorrosion, proposed in 1934 [52], was correct. As well, we have recently extended this work to demonstrate that aluminium participates in a similar process [53], and that pH plays a role in iron biocorrosion [54]. Although the phenomenon is easily demonstrated, it is unclear what quantitative role methanogens play in corrosion in the natural setting; clearly, non-biological metal corrosion is important in most ecosystems. In some environments, alkyl tin paints are used to control biofouling, a complex growth accumulation on surfaces, e.g. ship hulls and docks. Our studies show the most commonly used alkyl tin, tributyl tin, is relatively ineffective at inhibiting methanogens; indeed the pattern of inhibition of methanogens by a series of alkyl tins is opposite to the inhibition observed with marine organisms, and *Escherichia coli* [55,56]. If methanogens are to be controlled in these environments, it may be necessary to investigate other inhibitors.

Energy drain in ruminants

The animal rumen is a complex system with many microbial species present; feed material is broken down by microbes into fatty acids which are taken up by the ruminants as their major nutrients. In the same ecosystem, methanogens convert fermentation intermediates into methane [15,57–63]; this conversion of potential nutrients into methane, which then leaves the rumen intact by belching or other methods, represents lost nutrition. Loss of 5–15% of feed energy can occur due to methanogenesis.

Despite the huge economic benefit of understanding this system well, we know very little about the specific methanogens present in the rumen, although a variety have been isolated ([58–64] and references within), and much is known about the heterotrophic microbes present. Several hydrogen-dependent methanogens (e.g. *Methanobrevibacter* and *Methanobacterium* species) have been isolated from ruminants, and *Methanosarcina* species are thought to be important as direct competitors for acetate. However, neither variation of methanogen species with diet, age or drug treatment, nor the investigation of methanogen-specific inhibitors have been thoroughly studied; indeed, methanogen-specific inhibitors are largely unavailable.

One reason that we do not have more detailed information about rumen methanogens is that several microbial inhibitors, such as monensin and lasalocid, have been effective at both lowering the amount of methane emitted by ruminants, and at raising the weight gain and feed efficiency [15,58–60,62]; thus, conventional thought is that if currently used inhibitors work then there is no need to investigate further. It is thought that the inhibitors act via a variety of bacteria other than methanogens, leading to a higher propionate/acetate ratio, and greater protein availability. I contend that it is valuable to explore more completely the microbiology of the rumen, especially the methanogens, and to locate methanogen-specific inhibitors in an effort to enhance feed efficiency above current levels. It is also important to study a variety of different ruminants, rather than concentrate on a small range of species of young, well-fed cattle and sheep in the industrialized world.

Increased levels of methane in the atmosphere, leading to global warming

There is clear evidence from several types of data that greenhouse gases are increasing in our atmosphere ([14,65–68] and references within). One gas, carbon dioxide, arises largely from burning fossil fuels; another, methane, arises partly from non-biological sources such as natural gas-well leaks and natural seepage (approx. 0.14×10^9 tons/year), but more than half (approx. 0.4×10^9 tons/year) is biologically produced by methanogenic bacteria. Methane will have an even greater impact than carbon dioxide in the coming decades. It will cause temperature rises in several regions of the world; the exact magnitude and regional distribution is uncertain, but that it will occur is a scientifically sound conclusion. The temperature rise will cause agricultural disasters in some regions, but in some areas it may increase rainfall, with the potential for improved agricultural productivity.

Currently, methane flux into the atmosphere from bacteria is thought to arise mainly from tropical and temperate wetlands, ruminants, and rice paddies [14,65–68]. The ability of methanogens to compete effectively for substrates can be regulated by the availability of sulphate (which allows sulphate-reducing bacteria to out-compete methanogens), nitrate (which may lead to inhibition of methanogens [69], or may allow nitrate reducers to compete for substrates), and other factors. Water must be sufficiently abundant to maintain anaerobicity, and temperature can affect the rates of metabolism of all bacteria in the ecosystem. Owing to the action of methane-consuming bacteria, soils and aerobic portions of bodies of water are thought to be important in methane removal; ammonium can inhibit methane oxidation in some cases [70,71]. Thus, the net flux of methane into the atmosphere is regulated by a complex of features that influence the microbial ecology of the system.

There are two ecosystems which are both major producers of methane, and under control of man to some degree: rice paddies, and ruminants. For example, nitrogen-containing fertilizers could either inhibit methanogens (nitrate) or methane oxidizers (ammonium) when applied to rice paddies; drier temperate field soils may also be affected in a similar way. Also, our control of ruminant methane production could improve feed efficiency, and also decrease methane flux into the atmosphere. It is of interest to note that more than 30% of the world's cattle are in tropical and sub-tropical regions, and more than two-thirds of the rice is grown in these regions. It thus follows that the details of methane production via these sources should be studied in more detail, and the possible modification of agricultural practices to exert ecological control mechanisms should be examined carefully.

References

1. Wolfe, R. S. (1979) Antonie van Leeuwenhoek J. Microbiol. **45**, 353–364
2. Jones, W. J., Nagle, D. P. & Whitman, W. B. (1987) Microbiol. Rev. **51**, 135–177
3. Bhatnagar, L., Jain, M. K. & Zeikus, J. G. (1991) in Variations in Autotrophic Life, pp. 251–269, Academic Press, San Diego
4. DiMarco, A. A., Bobik, T. A. & Wolfe, R. S. (1990) Annu. Rev. Biochem. **59**, 335–394
5. Jarrell, K. F. & Kalmokoff, M. L. (1988) Can. J. Microbiol. **34**, 557–576
6. Sebacher, D. I., Harris, R. C., Bartlett, K. B., Sebacher, K. B. & Grice, S. S. (1986) Tellus **38B**, 1–10
7. Whalen, S. C. & Reeburg, W. S. (1988) Global Biogeochem. Cycles **2**, 399–408
8. Whalen, S. C. & Reeburg, W. S. (1990) Tellus **42B**, 237–249
9. Miller, T. L. & Wolin, M. J. (1986) Syst. Appl. Microbiol. **7**, 223–229
10. Daniels, L. (1984) Trends Biotechnol. **2**, 91–98
11. Sahm, H. (1984) in Advances in Biochemical Engineering/Biotechnology (Fiechter, A., ed.), pp. 83–116, Springer-Verlag, Berlin
12. Belay, N. & Daniels, L. (1988) Antonie van Leeuwenhoek J. Microbiol. **54**, 113–125
13. Daniels, L., Belay, N., Rajagopal, B. S. & Weimer, P. J. (1987) Science **237**, 509–511
14. Cicerone, R. J. & Oremland, R. S. (1988) Global Biogeochem. Cycles **2**, 299–327
15. Russell, J. B. & Strobel, H. J. (1989) Appl. Environ. Microbiol. **55**, 1–6
16. Stafford, D. A., Wheatly, B. I. & Hughes, D. E. (eds.) (1980) Anaerobic Digestion, Applied Science Pub., London
17. Hall, E. R. & Hobson, P. N. (eds.) (1988) Anaerobic Digestion 1988, Pergamon Press, Oxford
18. Lettinga, G., van Velsen, A. F. M., Hobma, S. W., de Zeeuw, W. & Klapwijk, A. (1980) Biotech. Bioeng. **22**, 699–734
19. Switzenbaum, M. S. (1983) Enzyme Microbiol. Technol. **5**, 242–250
20. Bungay, H. R. (1981) Energy, the Biomass Options, Wiley Interscience, New York
21. Lettinga, G., Hobma, S. W., Pol, L. W. H., de Zeeuw, W., de Jong, P., Grin, P. & Roersma, R. (1983) Water Sci. Technol. **15**, 177–195
22. Henze, M. & Harremoes, P. (1983) Water Sci. Technol. **15**, 1–101
23. Wise, D. L., Cooney, C. L. & Augenstein, D. C. (1978) Biotechnol. Bioeng. **20**, 1153–1172
24. Bower, E. J. & McCarty, P. L. (1983) Appl. Environ. Microbiol. **45**, 1286–1294

25. Egli, C., Tschan, T., Scholtz, R., Cook, A. M. & Leisinger, T. (1988) Appl. Environ. Microbiol. **54**, 2819–2824
26. Fathepure, B. Z. & Boyd, S. A. (1988) Appl. Environ. Microbiol. **54**, 2976–2980
27. Belay, N. & Daniels, L. (1987) Appl. Environ. Microbiol. **53**, 1604–1610
28. Vogel, T. M. & McCarty, P. L. (1985) Appl. Environ. Microbiol. **49**, 1080–1083
29. DiStefano, T. D., Gossett, J. M. & Zinder, S. (1991) Appl. Environ. Microbiol. **57**, 2287–2292
30. Van der Meijden, P., Heythuysen, H. J., Powels, A., Houwen, F., Van der Drift, C. & Vogels, G. D. (1983) Arch. Microbiol. **134**, 238–242
31. Van der Meijden, P., Jansen, L. P. J. M., Van der Drift, C. & Vogels, G. D. (1983) FEMS Microbiol. Lett. **19**, 247–251
32. Van der Meijden, P., Te Brommelstroet, B. W., Poirot, C. M., Van der Drift, C. & Vogels, G. D. (1984) J. Bacteriol. **160**, 629–635
33. Van der Meijden, P., Van der Lest, C., Van der Drift, C. & Vogels, G. D. (1984) Biochem. Biophys. Res. Commun. **118**, 760–766
34. Nagle, D. P. & Wolfe, R. S. (1983) Proc. Natl. Acad. Sci. U.S.A. **80**, 2151–2155
35. Gunsalus, R. P. & Wolfe, R. S. (1980) J. Biol. Chem. **255**, 1891–1895
36. Gunsalus, R. P., Rommesser, J. A. & Wolfe, R. S. (1978) Biochemistry **17**, 2374–2377
37. Wackett, L. P., Honek, J. F., Begley, T. P., Wallace, V., Orme-Johnson, W. H. & Walsh, C. T. (1987) Biochemistry **26**, 6012–6018
38. Mazumder, T. K., Nishio, N., Fukuzaki, S. & Nagai, S. (1987) Appl. Microbiol. Biotechnol. **26**, 511–516
39. Silveira, R. G., Nihsio, N. & Nagai, S. (1991) J. Ferm. Bioeng. **71**, 28–34
40. Koesnandar, Nishio, N. & Nagai, S. (1991) J. Ferm. Bioeng. **71**, 181–185
41. Stupperich, E., Steiner, I. & Eisinger, H. J. (1987) J. Bacteriol. **169**, 3076–3081
42. Stupperich, E. & Krautler, B. (1988) Arch. Microbiol. **149**, 268–271
43. Hatanaka, H., Wang, E., Taniguchi, M., Iijima, S. & Kobayashi, T. (1988) Appl. Microbiol. Biotechnol. **27**, 470–473
44. Crueger, W. & Crueger, A. (1989) Biotechnology, Sinauer Assoc., Sunderland, M.A.
45. Daniels, L., Bakheit, N. & Harmon, K. (1985) Syst. Appl. Microbiol. **6**, 12–17
46. Eker, A. P. M., Pol, A., Van der Meijden, P. & Vogels, G. D. (1980) FEMS Microbiol. Lett. **8**, 161–165
47. McCormick, J. R. D. & Morton, G. O. (1982) J. Am. Chem. Soc. **104**, 4014–4015
48. Rhodes, P. M., Winskill, N., Friend, E. J. & Warrenn, M. (1981) J. Gen. Microbiol. **124**, 329–338
49. Kuo, M. S. T., Yurek, D. A., Coats, J. H. & Li, G. P. (1989) J. Antibiotics **42**, 475–478
50. Ashton, W. T., Brown, R. D., Jacobson, F. & Walsh, C. (1979) J. Am. Chem. Soc. **101**, 4419–4420
51. Jacobson, F. & Walsh, C. (1984) Biochemistry **23**, 979–988
52. Von Wolzgen Kuhr, C. A. H. & Van der Vlught, L. S. (1934) Water **18**, 147–165
53. Belay, N. & Daniels, L. (1990) Antonie van Leeuwenhoek J. Microbiol. **57**, 1–7
54. Boopathy, R. & Daniels, L. (1991) Appl. Environ. Microbiol. **57**, 2104–2108
55. Belay, N., Rajagopal, B. S. & Daniels, L. (1990) Curr. Microbiol. **20**, 329–334
56. Boopathy, R. & Daniels, L. (1991). Appl. Environ. Microbiol. **57**, 1189–1193
57. Hungate, R. E. (1966) The Rumen and Its Microbes, Academic Press, New York
58. Chen, M. & Wolin, M. J. (1979). Appl. Environ. Microbiol. **38**, 72–77
59. Thornton, J. H. & Owens, F. N. (1981) J. Animal Sci. **52**, 628–634
60. Wedegaertner, T. C. & Johnson, D. E. (1983) J. Animal Sci. **57**, 168–177
61. Pol, A. & Demeyer, D. I. (1988) Appl. Environ. Microbiol. **54**, 832–834

62. Wallace, R. J., Cheng, K. J. & Czerkawski, J. W. (1980) Appl. Environ. Microbiol. **40**, 672–674
63. Vicini, J. L., Brulla, W. J., Davis, C. L. & Bryant, M. P. (1987) Appl. Environ. Microbiol. **53**, 1273–1276
64. Mukhopadhyay, B., Purwantini, E., Conway de Macario, E. & Daniels, L. (1991) Curr. Microbiol. **23**, 165–173
65. Seiler, W. (1984) in Current Perspectives in Microbial Ecology (Klug, M. J. & Reddy, C. A., eds.), pp. 468–477, ASM, Washington DC
66. Ramanathan, V., Ciceron, R. J., Singh, H. B. & Kiehl, J. T. (1985) J. Geophys. Res. **90**, 5547–5566
67. Sheppard, J. C., Westberg, H., Hopper, J. F. & Ganesan, K. (1982) J. Geophys. Res. **87**, 1305–1312
68. Blake, D. R. & Rowland, F. S. (1988) Science **239**, 1129–1131
69. Montfort, D. O., Asher, R. A., Mays, E. L. & Tiedje, J. M. (1980) Appl. Environ. Microbiol. **39**, 284–286
70. Ferenci, T., Strom, T. & Quayle, J. R. (1975) J. Gen. Microbiol. **91**, 79–91
71. Steudler, P. A., Bowden, R. D., Meillo, J. M. & Aber, J. D. (1989) Nature (London) **341**, 314–316

Biochem. Soc. Symp. **58**, 195–207
Printed in Great Britain

Where next with the archaebacteria?

Otto Kandler

Botanical Institute, University of Munich, Germany

Introduction

The accumulating molecular and physiological data of 14 years of research triggered by Carl Woese's archaebacteria concept [1,2] have led to the proposal that living beings on this planet should be classified into three domains [3]; Archaea*, Bacteria*, and Eukarya. The numerous new findings and ideas reported during this symposium have again shown the uniqueness of the biochemical and molecular features of the archaea and thus support their separate grouping aside from the bacteria. However, most speakers have also concluded their talks emphasizing that many open questions still await clarification, and they have identified problems within their particular fields, which will, or at least should, be tackled next.

Perspectives in cell wall chemistry

Before discussing the main goals of future archaeal research, I should like to add briefly the recent development and the perspective of further research in my own particular field, the chemistry of the rigid cell wall sacculi of archaea. Actually, the lack of murein in archaea [4,5] on the one hand, and the occurrence of chemically diverse non-murein cell wall polymers in some groups of archaea [6–8] on the other, has been one of the first recognized examples of basic biochemical differences between the bacterial and the archaeal molecular phenotype [1,2,5,9], though the biosynthesis of methanochondroitin, the cell wall polymer of *Methanosarcina* [6], and that of the sulphated heteropolysaccharide found in the cell wall of *Halococcus* [7], have not yet been elucidated. Only for pseudomurein, the cell wall polymer of the Methanobacteriales, has a scheme of biosynthesis been proposed [10] on the basis of

* The recently proposed formal names Bacteria and Archaea [3] will be used throughout this paper instead of the vernacular names eubacteria and archaebacteria, respectively.

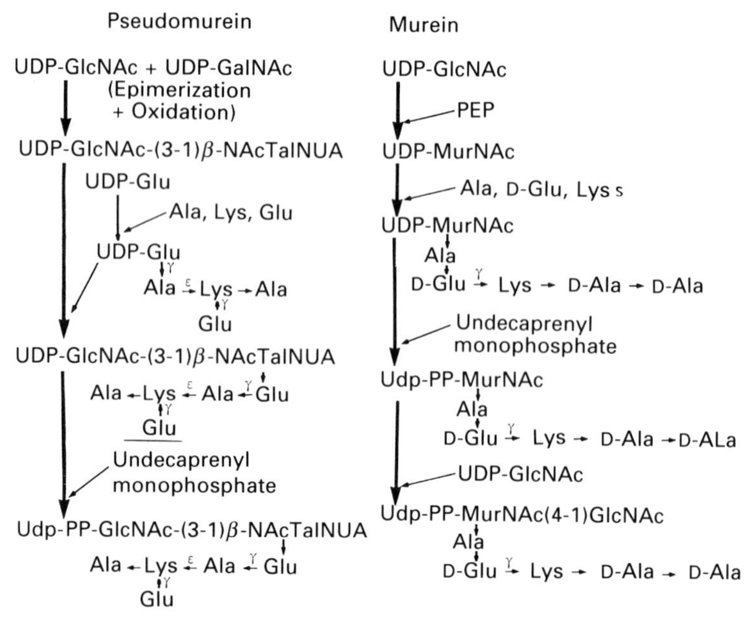

Fig. 1. Scheme of the first steps of biosynthesis of methanobacterial pseudomurein and bacterial murein. Abbreviations: TalNUA, talosamine uronic acid; UDP, uridine diphosphate; Udp, undecaprenyl; Mur, muramic acid. The Udp-activated disaccharide pentapeptides are transferred to the glycan backbone of the respective polymer. Cross-linking of the glycan strands is brought about by transpeptidation between adjacent peptide subunits and release of the terminal alanyl residue. In the case of pseudomurein the cross-link probably extends from the γ-carboxyl group of glutamic acid to the amino-group of lysine of an adjacent peptide subunit. (Adapted from [10]).

putative cell wall precursors recently isolated from *Methanobacterium thermoautotrophicum* [11,12]. The basic differences in the biosynthesis of murein and of pseudomurein (depicted in Fig. 1) corroborate our earlier assumption, on the basis of structural differences between the two polymers, that pseudomurein is not merely one of the many chemotypes of the bacterial murein [13], but a structure *sui generis* evolved independently within the methanobacterial lineage [5,8]. Among the various unusual reactions in the biosynthesis of pseudomurein, the formation of UDP-activated glutamic acid and UDP-activated di-, tri- and pentapeptides cayrring a UDP-residue directly linked to the N^{α}-amino group of a glutamic residue [11,12] is the most striking feature not found in nature before. The elucidation of the reaction mechanisms and the enzymes involved in the biosynthesis of pseudomurein is obviously one of the most urgent tasks in this field.

In addition to studies on the biosynthesis of known polymers, the search for new types of cell wall polymers in recent and forthcoming isolates of archaea is also promising, as shown by the recent discovery of a new cell wall polymer in *Natronococcus occultus*, an alkaliphilic halophile isolated from a natron lake in Egypt and first described in 1984 [14,15]. The polymer contains glucuronic acid, glucosamine and glutamic acid in a molar ratio of 1:1:1 (Kandler, O., Tindall, H. N., Fiedler, F., König, H., unpublished work). Such a composition is not known from any other cell wall polymer. Thus, the cell wall of *N. occultus* is an additional example of the independent invention of a variety of chemically different cell wall polymers in various lineages of the domain Archaea.

Considerable progress has also been achieved in elucidating the chemical and three-dimensional molecular structures of proteinaceous cell envelopes and sheaths of archae not possessing rigid cell walls (cf. [16,17]). A tremendous diversification of structures has been found, and the discovery of further unique features of envelope polymers is to be expected.

Main goals of future archaeal research

Future achievements along the various lines of biochemical research discussed during this meeting, together with progress in paleontology and other earth sciences, will result in a better understanding of (*a*) the interdependence between archaea and the evolution of the environment, (*b*) their alleged connection with the origin of life, and (*c*) their place in a natural (genealogical) system of organisms, the three major topics of broad interdisciplinary interest.

(A) The interdependence of archaeal and environmental evolution

Until recently, the first organisms on earth were assumed to have been mesophilic heterotrophs, thriving in a prebiotic broth, whereas a thermophilic origin of life was thought to be most unlikely for two reasons [18]: 'If early life had been subjected to selection by high temperatures, the argument goes, temperature-resistant enzymes would be far more widespread than they are. Moreover, if the earliest forms of life evolved in a high-temperature regime, one might expect to see a phylogenetic correlation between degree of 'primitiveness' and thermophily among extant microbes, but this is not the case.'

(1) Hyperthermophily, a relic of Archaean environmental conditions? The scene has been changing radically with the recent discovery of an almost logarithmically increasing number of mostly chemolithoautotrophic hyperthermophiles (Fig. 2; Table 1). These organisms grow optimally in the range 75–110 °C but do not grow at all at mesophilic temperatures. However, they are able to survive for years at low temperatures (1–9 °C) and thus can be distributed by ocean currents over long distances when they become released from their volcanic niches and dispersed in the ocean, for instance by the eruption of volcanoes [20]. All the hyperthermophiles exhibit the expected phylogenetic correlation between 'primitiveness' and thermophily, since their lineages represent the deepest branchings close to the common root of the phylogenetic tree based on 16S rRNA sequence comparison (Fig. 3). Most of the hyperthermophiles belong to the domain Archaea. To this day the kingdom of Crenarchaeota consists exclusively of hyperthermophiles, whereas in the kingdom Euryarchaeota only the lower branches are formed by hyperthermophiles, while the more distant branches are formed by mesophilic methanogens and halophiles. The domain Bacteria has only two lineages of hyperthermophiles, that of the heterotrophic Thermotogales and that of the two

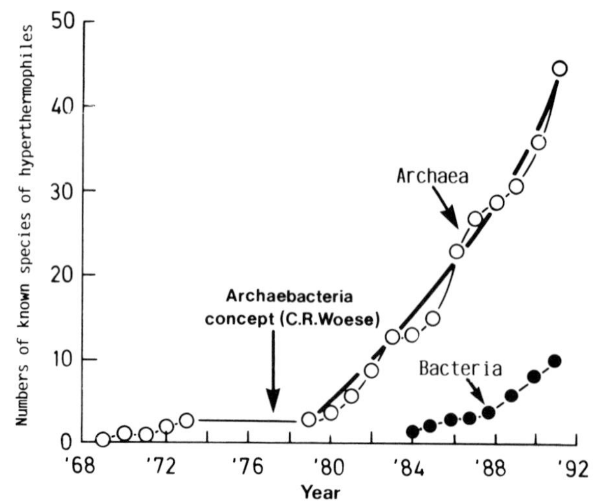

Fig. 2. Increase in the total number of known hyperthermophilic species after C. R. Woese's publication of the archaebacteria concept [1,2].

related H_2/O_2 chemolithoautotrophic genera *Hydrogenobacter* [21,22] and *Aquifex* [23]. They form the two deepest branches. Remarkably, all the moderately thermophilic species of bacteria, which grow from the mesophilic temperature range to up to about 80 °C, are scattered among close mesophilic relatives. This indicates that moderate thermophily is not a primitive, but rather a derived, character. We may call this a 'secondary thermophily', acquired by genetic adaptation of mesophiles to moderately heated, mostly continental biotopes, such as the effluent of hot wells or geysers, self-heated biomass, etc. 'Primary hyperthermophily' by contrast, is probably a relic of the high-temperature regime of the primitive earth.

(2) Primitive autotrophic pathways, a characteristic of archaea and early bacteria. Most of the hyperthermophilic archaea and bacteria are obligately or facultatively $H_2/S°$, H_2/CO_2 or H_2/O_2 chemolithoautotrophic, but none is photoautotrophic (Table 1). This suggests a priority of chemoautotrophy over photoautotrophy. Unlike the majority of the mesophilic or moderately thermophilic autotrophic bacteria, the hyperthermophiles, bacterial and archaeal alike, do not use the Calvin–Benson cycle for CO_2 assimilation. Instead, they use various modifications of the reductive citric acid cycle or of the reductive acetyl-CoA pathway ([24], G. Fuchs *et al.*, pp. 23–39 in this book), the exclusive pathways for carbon assimilation used by the archaea. It is most likely that these pathways have preceded the much more sophisticated Calvin–Benson cycle. This is supported by the finding that the two deep-rooted photosynthetic bacterial lineages, the green sulphur bacteria and the green non-sulphur bacteria, do not use the Calvin–Benson cycle, but instead a reductive citric acid cycle or the still largely unexplored reductive malonyl-CoA pathway which may well be a derivative of the reductive citric acid cycle (G. Fuchs *et al.*, pp. 23–39 in this book). By contrast, the Calvin–Benson cycle and its coupling with oxygenic and anoxygenic photosystems or various chemolithotrophic mechanisms is restricted to several phylogenetically derived mesophilic bacterial

Table. 1. Characteristics of hyperthermophilic genera (growth at > 80 °C). (Modified after [19]).

Genus	$T_{max.}$ (°C)	Aerobe or anaerobe	Growth: Hetero-trophic	Growth: Auto-trophic	Electron donor	Electron acceptors
Archaea						
Methanothermus	97	an	−	+	H_2	CO_2
Methanococcus	86	an	−	+	H_2	CO_2
Methanopyrus	110	an	−	+	H_2	CO_2
Stygiolobus	88	an	−	+	H_2	S°
Acidianus	95	ae/an	−	+	S°; H_2; Me-S	O_2; S°
Desulfurolobus	95	ae/an	−	+	S°; H_2	O_2; S°
Sulfolobus	85	ae	+	+	S°; H_2S; org.	O_2
Metallosphaera	80	ae	+	+	S°; Me-S	O_2
Thermoproteus	97	an	+	+	H_2; org.	S°; $S_2O_3^{2-}$
Pyrobaculum	103	an	+	+	H_2; org.	S°; $S_2O_3^{2-}$
Thermofilum	95	an	+	−	org.	S°
Desulfurococcus	95	an	+	−	org.	S°
Staphylothermus	98	an	+	−	org.	S°
Pyrodictium	110	an	+	+	H_2; org.	S°; ($S_2O_3^{2-}$)
Hyperthermus	107	an	+	−	org.	S°; ferm.
Thermodiscus	88	an	+	−	org.	S°
Thermococcus	97	an	+	−	org.	S°; ferm.
Pyrococcus	103	an	+	−	org.	S°; ferm.
Archaeoglobus	95	an	+	+	H_2; org.	SO_4^{2-}; $S_2O_3^{2-}$
Bacteria						
Hydrogenobacter	77	ae	−	+	H_2; S°; $S_2O_3^{2-}$	O_2
Aquifex	93	(an)	−	+	H_2; S°; $S_2O_3^{2-}$	O_2; NO_3
Fervidobacterium	80	an	+	−	org.	ferm.
Thermotoga	90	an	+	−	org.	ferm.

Abbreviations used: T_{max}., maximum growth temperature; an, anaerobe; ae, aerobe; (an), microaerophilic; org., organic compounds; Me-S, sulphidic ores; ferm., electron transfer by fermentation.

branches. This means that the Calvin–Benson cycle may be a phylogenetically rather late invention, an autapomorphy of the domain Bacteria, whereas the reductive citric acid cycle and the reductive acetyl-CoA pathway, primarily coupled with H_2-dependent chemolithoautotrophy, may belong to the common heritage of both domains. However, for further support of this suggestion it will have to be shown that the respective enzymes in the two domains are homologous.

(3) Adaptation to a changing world. The new qualities of the environment created by the gradual cooling of the earth and by the invention of oxygenic photosynthesis were maximally exploited by bacteria and eukarya as evidenced by a tremendous evolutionary radiation within these two domains and a virtually complete colonization of the numerous new biotopes. The archaea were much less progressive. Within the euryarchaeotes there was some adaptation to microaerophilic and moderate thermophilic growth in the case of *Thermoplasma*, and to mesophilic growth

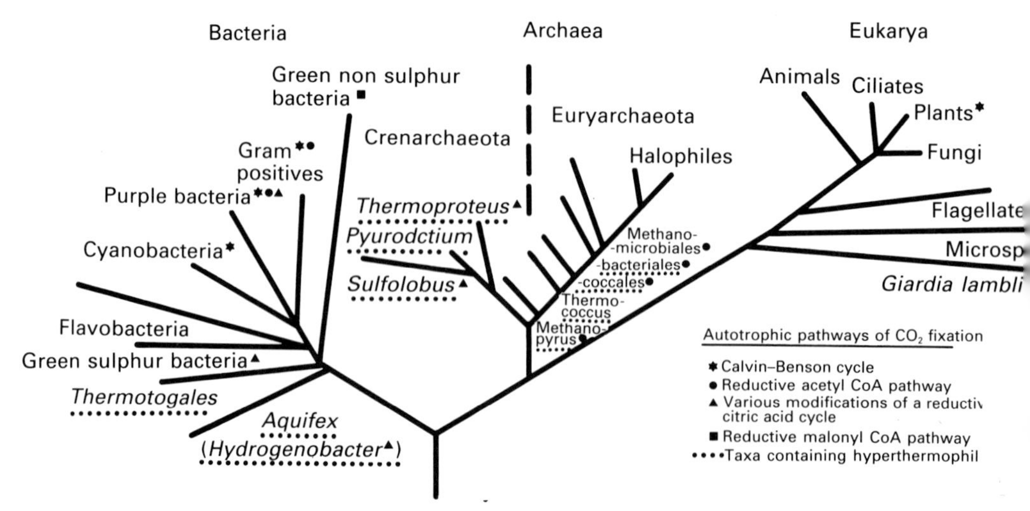

Fig. 3. Universal phylogenetic tree (3) showing the division of organisms into three domains and the subdivision of the domain Archaea into two kingdoms (kingdoms within the two other domains remain to be defined). Branching order and branch lengths are based upon rRNA sequence comparisons. The position of the root was determined by comparing known sequences of paralogous genes that diverged from each other before the three primary lineages emerged from their ancestral condition. The originally published tree [3] is modified by the addition of the branches for three recently sequenced organisms: *Giardia lamblia* [59], *Methanopyrus kandleri* [60], *Aquifex pyrophilus* [62]. The 16S rRNA of a *Hydrogenobacter* strain related to *Aquifex pyrophilus* is only partially sequenced [61].

and to a wider range of substrates in some genera of methanogens. However, a breakthrough to aerobic and mesophilic life was only achieved by the halophiles which colonized the evaporating continental (salt-, soda-) lakes. They even seem to be on their way to evolving their own type of photosynthesis with retinal instead of chlorophyll as photoreceptor. So far, only some species exhibit photoorganotrophic growth [25], but no known halophile is capable of photosynthetic CO_2 assimilation. The crenarchaeotes remained completely caught in their heated niche, and they remained dependent on H_2 and sulphur. Only a few branches have adapted to aerobic life, but even these remained confined to high temperatures and to sulphur metabolism. One wonders if this picture will change when a much larger number of crenarchaeal taxa become known. For the time being, it seems reasonable to conjecture that the extant hyperthermophilic chemolithoautotrophic or sulphur-respiring organotrophic archaea, found in the volcanic areas all over the globe [20,22,23,26–28] represent the least modified descendants of early organisms which might have operated the very first biological carbon cycle on earth.

(4) Hyperthermophiles, a component of the earliest biological cycle of matter? Unfortunately, no meaningful estimate of the total biomass produced by hyperthermophiles in their natural biotopes is available as yet. Most likely, the biomass of hyperthermophiles represents only a very small portion of the extant total biomass, whereas its portion was probably much higher in earlier geological periods. It may have been the dominant or even the only biomass in the dawn of life during the latest Hadean* and the early Archean* time (about 4.0–3.8 Ga), when the benthic region of the ocean was still heated. In this respect it is noteworthy that the ratio between organic carbon ($C_{org.}$) and carbonate carbon ($C_{carb.}$) of 20:80 % and the dilution of the heavy carbon isotope in $C_{org.}$ as indicated by a $\delta^{13}C_{org.}$** value of $-25‰$ were found to be virtually constant in rocks from the various periods throughout the earth's history, even in the 3.8 Ga-old rocks of the Isua series [30,31]. This long-range near-constancy of important parameters of the carbon balance indicates that the rate of biomass production was constant at least for 3.8 Ga in spite of the drastic changes of both the environment and the organization of life. Thus, it was suggested '... that the $C_{org.}/C_{carb.}$ ratio in the atmosphere–ocean–crust system had been held constant throughout the ages by a roughly uniform rate of phosphorus cycling through the exogenic (superficial) compartment.' [32].

However, nothing is known as yet about the types of organisms which may have produced the assumed large amount of $C_{org.}$ in times before the period of stromatolites which are likely indicators for the existence of photoautotrophs. Usually it is assumed that the $C_{org.}$ found in the 3.8 Ga-old rocks of the Isua series may have originated from biological CO_2 fixation [31], probably by photoautotrophs [32]. However, no unequivocally identified microfossils are known to be present in these rocks [34,35], and thus this assumption rests merely on the isotope data. Unfortunately, the resolution provided by the isotope data is too low to distinguish between different pathways of CO_2 assimilation, because the $\delta^{13}C_{org.}$ values for the various autotrophic pathways differ within a wide range from about 8–12 % in organisms using the reductive citric acid cycle (e.g. $H_2/S°$ autotrophs) to up to 34–40 % in organisms using the reductive acetyl-CoA pathway [e.g. methanogens (Fuchs *et al.* pp. 23–39 in this book)]. Hence, $\delta^{13}C_{org.}$ values of about $-25‰$ are not necessarily indicative of photosynthetic CO_2 assimilation via the Calvin–Benson cycle. They may well represent mean values resulting from concomitant CO_2 assimilation by hyperthermophilic $H_2/S°$ and H_2/CO_2 chemolithoautotrophs via a reductive citric acid cycle and a reductive acetyl-CoA pathway, respectively. If such mixed microbial populations, together with hyperthermophilic sulphur-respiring organotrophic anaerobes as decomposers, formed the earliest microbial mats, sulphur may have been the link between organismic metabolism and geochemical processes in which sulphur played a prominent role. In addition to small portions of elemental sulphur, both reduced (sulphide) and oxidized (sulphate) species of sulphur were abundant in silicate melts, in the gas phase and in a large variety of minerals.

* Terminology in accordance with [29].

** $\delta^{13}C = \left[\frac{(^{13}C/^{12}C)_{sa}}{(^{13}C/^{12}C)_{st}} - 1\right] \times 10^3$ [‰, PDB] where sa = sample, st = standard, commonly the Peedee belemnite standard (PDB) in which $^{12}C/^{13}C = 88,99$ and $\delta^{13}C$ is defined as equaling 0.00‰ [30].

(B) Geochemical processes, sources of energy and reducing power for life?

The most far reaching scenario of an intimate interaction between sulphur chemistry and life has been developed by Wächtershäuser [35–37]. He sees pyrite formation [equation (**I**)] as the first source of energy and reducing power for CO_2 reduction [equation (**II**)] and the origin of life.

$$\text{(I)} \quad FeS + H_2S \text{ (aqueous)} \rightarrow FeS_2 + H_2 \ \Delta G^\circ = -41.9 \text{ kJ/mol}$$

$$\text{(II)} \quad CO_2\text{(aqueous)} + H_2 \rightarrow HCOOH \text{ (aqueous)} \ \Delta G^\circ = +30.2 \text{ kJ/mol}$$

$$\text{(III)} \quad FeS + H_2S \text{ (aqueous)} + CO_2 \text{ (aqueous)} \rightarrow FeS_2 + H_2O + HCOOH \ \Delta G^\circ = -11.7 \text{ kJ/mol}$$

The overall reaction [equation (**III**)] is thought to be favoured by the insolubility of pyrite even at low pH. However, only the reaction according to equation (**I**), which is enhanced at elevated temperatures, could be experimentally verified [38] so far.

In Wächtershäuser's scenario, H_2S is seen as a precursor of all biomolecules with catalytic sulphydryl groups, and pyrite as a precursor of ferredoxin and iron–sulphur clusters [39]. In his eyes, the 'world of life' has been from its very beginning and is still today 'an iron–sulphur world'. The hyperthermophilic H_2/S° chemolithoautotrophic crenarchaeotes may thus be considered as extant chief witnesses of such a scenario.

A model for a similar, although less intimate, interaction between geochemical processes and H_2-utilizing organisms is the coupling of iron oxidation (corrosion) under anaerobic conditions [equation (**IV**)] with the growth of methanogens [equation (**V**)] by a 'microbially assisted cathodic depolarization of Fe°' [40], which resembles to some extent the interspecies hydrogen transfer between fatty-acid-oxidizing bacteria and methanogens in anaerobic syntrophic cultures [41].

$$\text{(IV)} \quad 8\ H^+ + 4\ Fe^\circ + CO_2 \xrightarrow{\text{methanogens}} CH_4 + 4\ Fe^{2+} + 2\ H_2O \ \Delta G^\circ = -136 \text{ kJ/mol}$$

$$\text{(V)} \quad 4\ H_2 + CO_2 \xrightarrow{\text{methanogens}} CH_4 + 2\ H_2O \ \Delta G^\circ = -139 \text{ kJ/mol}$$

However, a process requiring metallic iron does not meet the natural conditions, since the occurrence of metallic iron or that of a suitable equivalent in rocks and non-anthropogenic sediments seems unlikely, unless it is assumed that the formation of the iron core of the earth was not yet completed when liquid water first appeared. Nevertheless, we should keep an open mind for any kind of as yet undiscovered growth on the basis of interactions between H_2-dependent chemolithoautotrophs and geochemical processes, which may occur in the particular volcanic (primordial) environments of such organisms.

Talking about the far past, we should be aware that such scenarios about the origin of life are castles in the air and that the alleged trails to those castles are often trails to nowhere. However, among the many alleged imaginary castles we can think of, one of them should finally turn out to be as real as the origin of life has been. If we never try, we never will arrive. But we should not continue in the same direction, when the alleged signposts vanish.

(C) The place of the Archaea in a natural (genealogical) system of organisms

The biologist's innate longing for a universal order of organisms was vividly expressed as early as 1570 by Lobel. In the preface of his herbal (quoted by Arber [42]) he writes: 'For thus in an order, than which nothing more beautiful exists in the heavens or in the mind of a wise man, things which are far and widely different become, as it were, one thing.'

(1) Towards a natural (genealogical) system. While Lobel had in his mind an order based on phenetic similarity reflecting the plan of the creator, Darwin transformed the desire for a universal order into a new perspective and gave a directive on how to build a natural universal system. He stated: 'I believe that the arrangement of the groups within each class in due subordination and relation to other groups, must be strictly genealogical in order to be natural,...' [44].

The realization of Darwin's directive by phenetic analyses of extant and fossil specimens has been successful within groups of higher organisms, but became extremely difficult in unicellular protists, and is virtually impossible in the case of prokaryotes. Nevertheless, there have been numerous attempts to design universal systems since Haeckel published his 'Monophyletischer Stammbaum' in 1866 [44]. This very first universal tree and all the others proposed since are based on combinations of alleged genealogic relationships deduced from phenetic similarity and levels of organizations (grades). Strategies based on such combinations are sometimes called 'evolutionary classification' [45,46], a term implying that such classifications are true geneaological systems. However, it is only the perception that semantophoretic polymers (DNA, RNA, proteins) represent molecular chronometers [47] (recording the phylogenetic events throughout the organisms' history) that has enabled us to establish phylogenetic relationships, even among the most phenetically diverse organisms, and to construct a natural (genealogical) universal system.

Among the various semantophoretic polymers, 16S rRNA has turned out to be the most useful molecular chronometer [48]. The phylogenetic tree based on sequence comparisons of 16S rRNAs of organisms from all presently recognized kingdoms (Fig. 3) show three clearly separated genealogical groupings. As expected, all eukaryotes, including the so-called 'protists', are assembled within one monophyletic grouping, whereas the prokaryotes are segregated into two monophyletic groupings, which were initially called 'eubacteria' and 'archaebacteria' [2,9].

(2) The three domain system. This tri-partite living world, originally derived from a relatively few 16S rRNA sequences, is today corroborated by signature analysis based on a large collection of 16S rRNA sequences from more than 500 species [49], and by sequence comparison of additional but different semantophoretic molecules, for instance 23S rRNA [50], protein elongation factors [50–52], ATPases [50–53] and ribosomal protein [54]. It is also supported by a large number of autapomorphic molecular, biochemical and physiological features found in the three groupings. Therefore, we have proposed the new 'three domain system' (Fig. 3) which reflects the underlying phylogenetic grouping. We also have replaced the common name 'archaebacteria' by the formal name 'archaea' to avoid further confusion between the two prokaryotic domains.

The segregation of the prokaryotes into two domains equally ranked with the domain of the eukaryotes at the highest hierarchical level has caused considerable uproar among systematists. Some of them reject the 'three domain system' and see 'kingdoms in turmoil' [55], or argue in favour of a 'classification ... based on the traditional principles of classification which biology shares with all fields in which items are classified, as are books in a library or goods in a warehouse.' [56]. Although these authors do not question the usefulness of semantophoretic polymers as molecular chronometers to establish phylogenetic relationships among organisms, they propose alternative universal systems of organisms [55,57] which are dominated by phenetic similarity. Their systems still reflect the traditional prokaryote/eukaryote dichotomy, the most conspicuous phenetic distinction among organisms at the level of cell organization, which, however, lacks molecular phylogenetic support.

(3) The roots. The universal trees based on 16S rRNA sequence comparison remained unrooted because, until recently, no outgroup for rooting was available [48]. However, comparison of the few known, recently sequenced paralogous genes, that diverged from each other before the three primary lineages emerged from their common ancestral condition [51–53], now allows a rooting of the universal tree (Fig. 3), since one set of the aboriginally duplicated genes can be used as an outgroup for the other. The rooting obtained thereby [3] indicates that the bacterial lineage was the first to emerge from the ancestral condition, whereas the archaeal and the eukaryal lineages diverged later. Judged from the characteristics of the known phenotypes of the three lineages, the archaea have remained the most metabolically primitive (Table 1), being well suited to thrive under the above mentioned Archaean environmental conditions. On the other hand, the hyperthermophilic H_2/O_2 chemolithotrophic genera *Aquifex* and *Hydrogenobacter*, which form the deepest branch in the bacterial lineage, depend on free O_2 which, most likely, has not been available in the Archean environment, although there may have been local niches having low levels of O_2. Therefore, the two genera may be considered to represent the later stages of evolutionary adaptation of early bacteria to an aerobic environment, and thus resemble the facultative aerobic archaeal genera *Acidianus* and *Desulfurolobus* which grow either by the 'modern' S°/O_2- or by the 'archaic' H_2/S°-chemolithoautotrophy. However, the assumed ancestors of the partially O_2-adapted archaeal genera may have survived, while the respective anaerobic H_2/S° chemolithoautotrophic ancestors of the bacterial genera *Aquifex* and *Hydrogenobacter* may have become extinct or remained so far undiscovered. Considering the steep rise in the number of recently discovered hyperthermophilic prokaryotes (Fig. 2), we hopefully may see missing links between the archaeal and bacterial chemolithoautotrophic phenotypes in the near future.

The early phenotypes of the eukaryal lineage are even more obscure than those of the bacterial lineage. Unicellular organisms without mitochondria and an endoplasmic reticulum represent the deepest known branches in the eukaryal lineage. This indicates that these branchings may have occurred before the introduction of mitochondria by endosymbiosis. However, even these deepest branchings are separated from the origin of the eukaryal lineage by a very long (rRNA) sequence distance, suggesting that we may still be missing the bulk of the primitive early forms of eukaryal life. Obviously a search is urgently needed for more primitive, free-living forms among the so-called 'Archaezoa' [58] using new enrichment and isolation strategies including anaerobic and thermophilic conditions. The isolation and analysis of ever more phylogenetically ancient phenotypes and new Archean microfossils of each of the three domains will further elucidate the evolutionary steps which led from the common ancestral condition to the segregation of the three lineages (domains) representing the extant life on this planet. Only then will Lobel's desire be fulfilled: to see 'things which are far and widely different become, as it were, one thing.'

References

1. Woese, C. R. & Fox, G. E. (1977) Proc. Natl. Acad. Sci. U.S.A. **74**, 5088–5090
2. Woese, C. R., Magrum, L. J. & Fox, G. E. (1978) Mol. Evol. **11**, 245–252
3. Woese, C. R., Kandler, O. & Wheelis, M. L. (1990) Proc. Natl. Acad. Sci. U.S.A. **87**, 4576–4579
4. Kandler, O. & Hippe, H. (1977) Arch. Mikrobiol. **113**, 57–60
5. Kandler, O. (1979) Naturwissenschaften **66**, 95–105
6. Kreisl, P. & Kandler, O. (1986) Syst. Appl. Microbiol. **7**, 293–299
7. Schleifer, K. H., Steber, J. & Mayer, H. (1982) Zentrabl. Bakteriol. Hyg. I. Abt. Orig. C **3**, 171–178
8. Kandler, O. (1982) Zentralbl. Bakteriol. Hyg. I, Abt. Orig. C **3**, 149–160
9. Woese, C. R. (1982) Zentralbl. Bakteriol. Hyg. I, Abt. Org. C **3**, 1–17
10. Hartmann, E. & König, H. (1990) Naturwissenschaften **77**, 472–475
11. König, H., Kandler, O. & Hammes, W. (1989) Can. J. Microbiol. **35**, 176–181
12. Hartmann, E., König, H., Kandler, O. & Hammes, W. (1990) FEMS Microbiol. Lett. **69**, 271–276
13. Schleifer, K. H. & Kandler, O. (1972) Bacteriol. Rev. **36**, 407–477
14. Tindall, H. N., Roos, M. & Grant, W. D. (1984) Syst. Appl. Microbiol. **5**, 41–57
15. Grant, W. D. & Larsen, H. (1989) in Bergey's Manual of Systematic Bacteriology (Staley, J. T., Bryant, M. P., Pfennig, N. & Holt, J. G., eds.), vol. 3, pp. 2216–2219, Williams & Wilkins, Baltimore
16. Kandler, O. & König, H. (1985) in The Bacteria Archaebacteria (Woese, C. R. & Wolfe, R. S., eds.) vol. 8, pp. 413–458, Academic Press, Inc., Orlando, San Diego, New York, London, Toronto, Montreal, Sydney, Tokyo
17. Kandler, O. & König, H. (1992) in The Biochemistry of Archaea (Archaebacteria) (Kates, M., Kushner, D. J. & Matheson, A. T., eds.), Elsevier, Amsterdam, in the press,
18. Walker, J. C. G., Klein, C., Schidlowski, M., Schopf, J. W., Stevenson, J. & Walter, M. R. (1983) in Earth's Earliest Biosphere—its Origin and Evolution (Schopf, J. W., ed.), pp. 51–92, Princeton University Press, Princeton, New Jersey
19. Stetter, K. O., Fiala, G., Huber, G., Huber, R. & Segerer, A. (1990) FEMS Microbiol. Rev. **75**, 117–124
20. Huber, R., Stoffers, P., Cheminee, J. L., Richnow, H. H. & Stetter, K. O. (1990) Nature (London) **345**, 179–182
21. Kawasumi, T., Igarashi, Y., Kodama, T. & Minoda, Y. (1984) Int. J. Syst. Bacteriol. **34**, 5–10
22. Nishihara, H., Igarashi, Y. & Kodama, T. (1990) Arch. Microbiol. **153**, 294–298
23. Huber, R., Vilharm, Th., Huber, D., Rachel, R., Trincone, A., Burggraf, S., Rockinger, I., König, H., Fricke, H. & Stetter, K. O. (1992) Syst. Appl. Microbiol., in the press
24. Fuchs, G. (1989) in Autotrophic Bacteria (Schlegel, H. G. & Bowien, B., eds.), pp. 365–382, Science Tech. Publishers, Madison WI, and Springer, Berlin
25. Oesterhelt, D. & Krippahl, G. (1983) Ann. Microbiol. (Paris) **134B**, 137–150

26. Zillig, W., Holz, I., Janecovic, D., Klenk, H. P., Imsel, E., Trent, J., Wunderl, S., Forjatz, V. H., Coutinho, R. & Ferreira, T. (1990) J. Bacteriol. **172**, 3959–3965
27. Kurr, M., Huber, R., König, H., Jannasch, H. W., Fricke, H., Tincone, A., Kristjansson, J. K. & Stetter, K. O. (1991). Arch. Microbiol. **156**, 239–247
28. Huber, G., Huber, R., Jones, B. E., Lauerer, G., Neuner, A., Segerer, A., Stetter, O. & Degens, E. T. (1991) Syst. Appl. Microbiol. **14**, 397–404
29. Ernst, W. G. (1983) in Earth's Earliest Biosphere—Its Origin and Evolution (Schopf, J. W., ed.), pp. 41–52, Princeton University Press, Princeton, New Jersey
30. Schidlowski, M., Hayes, J. M. & Kaplan, I. R. (1983) in Earth's Earliest Biosphere—Its Origin and Evolution (Schopf, J. W., ed.), pp. 149–186, Princeton University Press, Princeton, New Jersey
31. Schidlowski, M. (1988) Nature (London) **333**, 313–318
32. Schidlowski, M. (1987) Annu. Rev. Earth Planet Sci. **15**, 47–72
33. Schopf, J. W., Hayes, J. M. & Walter, R. (1983) in Earth's Earliest Biosphere—Its Origin and Evolution (Schopf, J. W., ed.), pp. 361–384, Princeton University Press, Princeton, New Jersey
34. Pflug, H. D. (1990) in Sonderveröffentlichungen Geologisches Institut der Universität zu Köln 70 (Festschrift Ulrich Jux), Köln, pp. 296–401
35. Wächtershäuser, G. (1988) Syst. Appl. Microbiol. **10**, 207–210
36. Wächtershäuser, G. (1988) Microbiol. Rev. **52**, 452–480
37. Wächtershäuser, G. (1990) Proc. Natl. Acad. Sci. U.S.A. **87**, 200–204
38. Drobner, E., Huber, H., Wächtershäuser, G., Rose, D. & Stetter, K. O. (1990) Nature (London) **346**, 742–744
39. Wächtershäuser, G. (1992) in Frontier of life, Proc. 3rd Recontres de Blois, in the press
40. Daniels, L., Belay, N., Rajagopal, B. S. & Weimer, P. J. (1987) Science **237**, 509–511
41. McInerney, M. J., Mackie, R. I. & Bryant, M. P. (1981) Appl. Environ. Microbiol. **41**, 826–828
42. Arber, A. (1938) Herbals, Their Origin and Evolution, 2nd edn., Cambridge University Press, Cambridge
43. Darwin, C. (1859) On the Origin of Species by Means of Natural Selection, or the Preservation of Favored Races in the Struggle for Life. John Murray, London
44. Haeckel, E. (1866) Generelle Morphologie der Oranismen, Verlag Georg Reimer, Berlin
45. Mayr, E. (1990) Verh. Dtsch. Zool. Ges. **83**, 263–276
46. Mayr, E. & Ashlock, P. (1990) Principles of Systematic Zoology, 2nd edn., McGraw-Hill, New York
47. Zuckerkandl, E. & Pauling, L. (1965) J. Theor. Biol. **8**, 357–366
48. Woese, C. R. (1987) Microbiol. Rev. **51**, 221–271
49. Winker, S. & Woese, C. R. (1991) Syst. Appl. Microbiol. **14**, 305–310
50. Schleifer, K. H. & Ludwig, W. (1989) in the Hierarchy of Life (Fernholm, B., Bremer, K. & Jörnwall, H., eds.), pp. 103–117, Exacta Medica, Amsterdam
51. Iwabe, N., Kuma, K., Hasegawa, M., Osawa, S. & Miyata, T. (1989) Proc. Natl. Acad. Sci. U.S.A. **86**, 9355–9359
52. Miyata, T., Iwabe, N., Kuma, K., Kawanishi, Y., Hasegawa, M., Kishino, H., Mukohata, Y., Ihara, K. & Osawa, (1991) in Evolution of Life (Osawa, S. & Honjo, T., eds.), pp. 337–351, Springer-Verlag, Tokyo, Berlin, Heidelberg, New York, London, Paris, Hong Kong, Barcelona
53. Gogarten, J. P., Kibak, H., Dittrich, P., Taiz, L., Bowman, E. J., Bowman, B. J., Manolson, M. F., Poole, R. J., Date, T., Oshima, T., Konishi, J., Denda, K. & Yoshida, M. (1989) Proc. Natl. Acad. Sci. U.S.A. **86**, 6661–6665

54. Auer, J., Spicker, G. & Böck, A. (1990) Syst. Appl. Microbiol. **13**, 354–360
55. Margulis, L. & Guerrero, R. (1991) New Sci. **129**, 46–50
56. Mayr, E. (1991) Nature (London) **353**, 122
57. Mayr, E. (1990) Nature (London) **348**, 491
58. Cavalier-Smith, T. (1991) in Evolution of Life (Osawa, S. & Honjo, T., eds.), pp. 271–304, Springer-Verlag, Tokyo, Berlin, Heidelberg, New York, London, Paris, Hong Kong, Barcelona
59. Sogin, M. L., Gunderson, J. H., Elwood, H. J., Alonos, R. A. & Peattie, D. A. (1989) Science **243**, 75–77
60. Burggraf, S., Stetter, K. O., Rouviere, P. & Woese, C. R. (1991) Syst. Appl. Microbiol. **14**, 346–351
61. Burggraf, S., Stetter, K. O., Olsen, G. J. & Woese, C. R. (1992) Syst. Appl. Microbiol. in the press

Subject Index